HOLOGRAMS

FOCAL PRESS LTD.
31 Fitzroy Square, London W1P 6BH

FOCAL PRESS INC
10 East 40th Street, New York, NY 10016

Associated companies
Pitman Publishing Pty Ltd., Melbourne
Pitman Publishing New Zealand Ltd., Wellington

Holograms

HOW TO MAKE AND DISPLAY THEM

Graham Saxby

Focal Press Limited, London

Focal Press Inc, New York

BL   British Library Cataloguing in Publication Data

Saxby, Graham
  Holograms.
  1. Holography
  I. Title
  774   QC449   79-42653

ISBN 0 240 51054 2

First edition 1980

Text set in 11/12 pt VIP Times, printed and bound
in Great Britain at The Pitman Press, Bath

Contents

Introduction

Only a few years ago this book could not have been written. One or two scientists, it is true, had succeeded in making some grainy, dark and somewhat blurred holograms. Few ordinary people had heard of holography, let alone seen examples of it; and among those who had, most believed that an understanding of holographic techniques would demand at least a higher degree in both physics and mathematics.

Since those early days there have been considerable advances in the technology of coherent light sources and of fine-grain photographic emulsions. In order to make holograms one no longer needs access to a large optical laboratory full of expensive equipment. Today, anyone prepared to take a little trouble can produce perfectly good holograms at home. The basic equipment needed is just a helium–neon laser (which can cost no more than a good-quality single-lens reflex camera), plus a few quite cheap optical components, and some holographic film. With a little more equipment you can make larger, more sophisticated holograms. You can produce transmission holograms to use as pictures with three dimensions, or reflection holograms to use as pendants or as table-top ornaments, needing only sunlight to produce a brilliant image.

Holography is basically simple – simpler in principle than photography. It requires no lens, but captures and encodes the information contained in the light waves coming from the object into what is called an interference pattern. When this pattern is recorded on film, and the developed film is illuminated, light waves similar to those originally encoded re-emerge. Thus when we look at a hologram we see what appears to be the object itself, in its original position, and in three dimensions.

People often tend to think of holography as just another, rather gimmicky, version of photography, and limited at that. But holography operates on principles entirely different from those of photography. Indeed, the very property of light – diffraction – which makes holography possible, is the one which limits the performance of photographic systems using lenses.

It is true that a hologram is made on a photographic emulsion, and that the end-product is a visible image; but there the resemblance ends. The way in which the information is coded is quite different, and the image we see is different too. A holographic image is much nearer to the real thing than a photographic image. Indeed, it can be argued that the experience of viewing a holographic image is identical with that of viewing the object itself.

If you intend to make your own holograms, you will no doubt be curious as to how they achieve their extraordinary three-dimensional quality. The underlying concepts are not difficult to grasp, nor even particularly novel. Indeed, the optical phenomena relevant to holography had all been fully described before the end of the nineteenth century, though it was as late as 1948 before anyone foresaw their practical application. In that year Dennis Gabor published a paper on the subject of improving image quality in electron microscopy. In this paper he suggested making use of these phenomena in an entirely new way. He coined the word 'hologram' to describe the recorded image, the prefix 'holo' (entire) indicating that the image-forming information was coded over the entire area of the photographic plate.

However, the practical exploitation of Gabor's ideas had to wait another 12 years, until the invention of the laser in 1960. It was only then that Emmett Leith and Juris Upatnieks were able to show that holography was indeed a practical possibility, and the importance of Gabor's work (for which he eventually received a Nobel prize) was appreciated. Since then, holography has moved out from the optical research laboratory into industry and commerce; and it has steadily become more feasible for the amateur to make holograms with no more difficulty than table-top photography.

Because holography was invented by physicists, its principles were from the beginning developed by means of mathematics rather than intuition, unlike those of photography, which began as a hobby for gentlemen. However, even for the mathematically well-equipped some of the underlying theory is heavy going; for this reason I have avoided mathematics (except for Appendix 1, which gives a mathematical explanation of why holograms work). However, those readers who do have a mathematical background, and feel cheated by the lack of rigour in the theoretical parts of this book, can find references in Appendix 5 which give the full treatment. Fortunately for the rest of us, it is not particularly difficult to understand the principles intuitively.

The main burden of this book is in Part 2, which follows through the process of making holograms step by step, beginning with comparatively small-scale images using a holographic layout which you can set up on a chipboard plank or paving stone. When you have acquired sufficient confidence, you can go on to the more ambitious setups using a sandbox, as described in the later chapters of Part 2.

Of course, with simple home-made equipment one cannot hope to produce the giant holograms displayed nowadays in lavish exhibitions, nor the full-colour – and even moving-image – holograms that are possible with sophisticated equipment. Part 2 also deals with these more advanced holograms.

As an amateur, you can produce holograms that are, in aesthetic terms, every bit as good as the metre-square giants of the exhibitions. After all, many of the greatest painters – and photographers, for that matter – have been miniaturists. What is more, your amateur status means that you can forget commercial considerations and give full rein to your creativity.

If you follow the instructions in Part 2 carefully, you should experience no special difficulties in making holograms. But you will be more likely to be successful, and certainly get more enjoyment out of it, if you know something of the principles involved. Part 1, introduces these.

As you are going to be using a laser, you will no doubt be interested to know how they work. Part 1 describes these principles of laser action in simple language. The most important quality of laser light is what is called 'coherence'. This quality is essential for making holograms, and Part 1 explains the meaning of the term, as well as describing the nature of electromagnetic radiation in general. Of course, you do not need to know what coherence means, nor to understand the principles of laser action, in order to make holograms; and, if you wish, you can defer the reading of Part 1 until later, or even miss it out altogether. But if you want to make really high-quality holograms there are certain restrictions to take into account. An understanding of the way a laser works, and the properties of the light it produces, will help you to appreciate the nature of these restrictions.

Holography has a number of applications in industry, and considerable potential in the fields of commerce and entertainment. It has uses in medicine, too, and in some quarters has found favour as a metaphor for describing cognitive functions of the brain such as perception, recognition and recall. It is also becoming accepted as an art medium in its own right.

The plates in this book show some of the things holography can do; but because they are photographs of holograms, they lack the one quality that above all gives the holographic image its uniqueness: its three-dimensional quality. If you are to appreciate the eerie beauty of a hologram, there is no substitute for the real thing.

Part 1
Principles of holography

Incoherent and coherent light

What is light?

At first glance this seems a straightforward question, like 'What is a table?' but it is not. To define a table is no problem; all we have to do is to select some objects of furniture that are generally agreed to be tables, and list the qualities they have in common. But we cannot define light in this way. Indeed, it has been argued that we cannot define light at all. What we can do is to describe the way light behaves, and produce some kind of metaphor or 'model' that represents the aspects of its behaviour that concern us. Thus for most of this book we shall be treating light as electromagnetic waves of the same kind as radio waves, only of shorter wavelength. This model is satisfactory for explaining all the principles of holography and the properties of holograms. However, it cannot adequately describe how light is generated, nor explain in what way laser light differs from light given by, say, a filament lamp. For this purpose we shall have to use a model that represents light as being generated in the form of large numbers of individual pulses of electromagnetic energy called 'photons'.

Electromagnetic waves

Let us look at the notion of 'electromagnetic waves'. You probably know that the passage of an electric current through a conductor creates a magnetic field round it. We can confirm this by placing a compass needle near the conductor: when we switch

13

on the current the needle is deflected. If we place a second compass needle a short distance farther away, it is also deflected, apparently simultaneously. However, this deflection is not in fact quite simultaneous: the second needle is deflected a tiny fraction of a second later than the first. This delay is because the magnetic field does not appear instantaneously throughout the whole of space. It is propagated outwards from the source at approximately 300 000 000 metres (186 000 miles) a second. This figure is usually written 3×10^8, where the figure 8 represents the number of zeros that come after the 3. We shall be using this method of notation throughout the book. It is known as *scientific notation*, and you will find a formal definition in the Glossary at the end of the book.

NOTE: When a definition or explanation of a technical or scientific term is to be found in the Glossary, this is indicated by its being printed in italics the first time it appears.

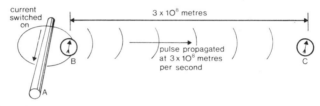

Propagation of an electromagnetic disturbance. **A** Conductor carrying an electric current. **B** Compass needle at conductor deflects at once **C** Compass needle 3×10^8 metres away deflects 1 second later.

As a magnetic field created in this way has an electrical field associated with it, we usually lump the two concepts together and talk about an *electromagnetic field;* and we call the propagation of the field through space *electromagnetic radiation*.

Let us think about our electrical conductor again, but this time let us imagine it carrying an AC mains current. This reverses in direction 100 times a second, so that each full cycle occurs 50 times a second (60 in the United States). We say that the *frequency* of the AC is 50 cycles per second, or 50 hertz (abbreviated Hz). Now, as these cycles are being continually propagated out into space at a fixed speed, each successive cycle will at any instant be spread over a given distance. We call this distance between successive 'wavecrests' the *wavelength*, and as the frequency becomes higher the wavelength becomes proportionately shorter. The wavelength of an oscillation of frequency 50 Hz works out at 6000 kilometres (about 3800 miles). It is often convenient to describe electromagnetic radiation in terms of wavelength rather than frequency.

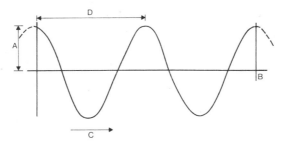

An electromagnetic wave. **A** Amplitude of wave is proportional to current flowing through source. **B** Frequency = number of wavecrests passing a point in 1 second. **C** Speed of propagation = 3×10^8 metres per second. **D** Wavelength (inversely related to frequency) = distance between successive crests.

The other quantity which concerns us is the *amplitude*, or height of the wavecrest above the mean. This depends on the magnitude of the current flowing in the conductor.

Visible electromagnetic waves

By using electronic devices called oscillators we can produce much higher frequencies than the 50 Hz of the AC mains, and correspondingly shorter wavelengths. The waves emitted are called radio waves. In a radio transmitter, a sound signal is coded onto the wave in the form of fluctuations in amplitude or frequency; the conductor that radiates the signal is called a transmitting aerial. At the receiving end a similar aerial, now called a receiving aerial, picks up the signal (i.e. the electromagnetic radiation induces a current to flow in it); and the receiver to which the aerial is connected decodes the signal back into sound.

If we could make our oscillator frequency high enough (about 5×10^{14} Hz) the radiated wave would be visible: it would be red light! Of course, we cannot make an electronic circuit that would operate at such a high frequency; but by using a device which stimulates atoms to emit waves at these frequencies we can produce either infrared or visible light as a continuous wave. The device is called a *laser*, and it takes its name from the initial letters that describe its method of operation: Light Amplification by Stimulated Emission of Radiation. It is this light source that we use for making holograms.

Conventional light sources

Lasers have been with us only a comparatively short time. Most conventional light sources are incandescent; that is, their operation depends upon their being at a temperature high enough for

15

some of their heat energy to be radiated away as visible light. Filament lamps are the most common incandescent source. Arc lamps, gas-mantle lamps and even candles are also incandescent sources.

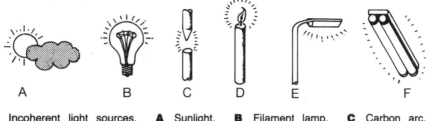

Incoherent light sources. **A** Sunlight. **B** Filament lamp. **C** Carbon arc. **D** Flame. **E** Gas discharge. **F** Fluorescent.

Another common light source is the gas discharge lamp, which derives its energy from an electric current passed through a gas in a glass tube. A proportion of the electrical energy is converted into light energy. Gas discharge lamps do not depend on high temperature for their operation, and are more efficient at converting electrical energy into light energy than are filament lamps.

In all types of light source it is the atoms of the substance that produce the light; and we shall see how this happens in the next chapter. However, the quality of the light differs from one source to another. One aspect of this is very important in holography. It is called *coherence*.

Waves and photons: incoherent light

In order to show how light from ordinary sources differs from laser light we shall have to introduce the concept of the *photon*. The atoms in a light-emitting substance do not radiate waves continuously: they emit discrete bursts of radiation (photons) at intervals. In an incandescent body, all the atoms are vibrating at random: photons of all frequencies are emitted in random directions, and at random times. The energy represented by a single photon is very small; but luminous bodies emit vast numbers of photons. The resultant of these many millions of photon 'events' is an apparently steady light output, in all directions, and with wavelengths ranged over the entire visible spectrum as well as invisible infrared and ultraviolet. This is called a continuous spectrum of radiation.

As an analogy, we may think of a crowd coming out of a

16

football ground, spreading in all directions. The people will, of course, be travelling at much the same speed; but individuals with shorter legs will be taking faster, shorter paces than those with longer legs (higher frequency, shorter wavelength) with the result that nobody is in step with any of his neighbours.

Light made up of photons of all frequencies is like this. It is said to be *incoherent*. All incandescent sources, as well as fluorescent lamps (which are a special kind of gas discharge lamp giving a continuous spectrum) emit incoherent light.

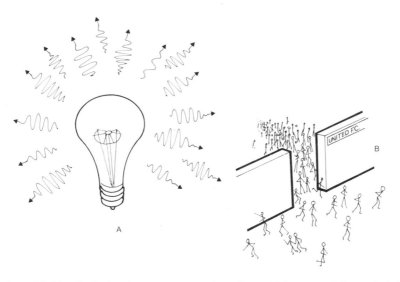

Incoherent light. **A** An incoherent source emits pulses of light energy (photons) of all frequencies, and in a random manner. **B** Analogy with a crowd emerging from a football match.

Partially coherent light

Gas discharge lamps emit their light only in narrow bands of wavelength, and are said to produce line spectra. Some, for example mercury vapour lamps, produce very few bright lines, so that by careful filtering we can eliminate all but one. Going back to our crowd analogy, we have something that now resembles a group of people, all of approximately the same height, emerging from a meeting. Their paces now have roughly the same 'wavelength', and thus they will remain more or less in step with one another for a short distance. We say that light from such a source is *partially coherent*. The distance over which the photons remain more or less in *phase* (i.e. in step) is called the *coherence length*.

17

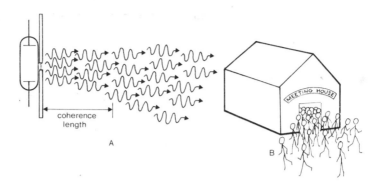

Partially coherent light. **A** Gas discharge lamp with slit and filter emits photons of a narrow range of frequencies which remain more or less in phase for a short distance. **B** Analogy with orderly dispersal of a meeting.

Gabor's first attempts at holography were made using the partially coherent light of a mercury vapour lamp: his 'object' was a photographic transparency which could afford to be little bigger than a pinhead, as the coherence length for mercury vapour light is less than a millimetre.

Although partially coherent light is not suitable for making holograms, at least on a large scale, it is nevertheless fairly satisfactory for viewing them.

Coherent light

Let us go back to our analogy. Let us take our group of similarly sized people, put them into uniform, arrange them in files on a parade ground, and march them up and down. They stay

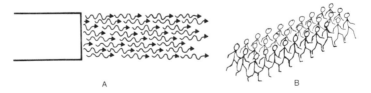

Fully coherent light. **A** Light from a laser has all photons of same frequency and all travelling in same direction. **B** Analogy with disciplined body of troops on the march.

together, and remain in step indefinitely. Their behaviour models that of *coherent* light, in which all the photons are of the same frequency and are in the same phase. Laser light provides a close

18

approximation to this, having a coherence length that may be many millions of wavelengths.

Temporal and spatial coherence

So far we have been discussing only *temporal coherence*; that is, the requirement for light to have a single frequency. A second requirement is *spatial coherence*. Ideally, all the photons should issue from a single point. If the source is an extended one, photons reaching a point from two separate points of origin will have travelled differing distances to get there, and be no longer in phase. This is an important restriction on partially coherent sources, as it means they can be used only behind a small aperture; this puts a serious limitation on an intensity already reduced by filtering. Now, a laser beam, in addition to having a high intensity, is also highly *collimated* (parallel). The photons therefore appear to have come from a single, very distant source. A laser beam is very nearly 100 per cent spatially coherent.

How a laser works

How atoms produce light

In order to understand how a laser works, we need to know something about the way a luminous substance produces light. As we have seen, incandescent lamps, gas discharge tubes and lasers emit light of differing qualities; but the photons all originate in a similar manner. They are emitted by the individual atoms that make up the substance. Let us examine this process, and see how we can control the usually random process of photon emission in order to produce coherent light.

An atom emits a photon only when its internal energy changes in a particular way; this happens as the result of a sudden rearrangement of its structure from a less stable to a more stable form. To see how this works we need to look more closely at this structure.

What is the structure of an atom? The question is by no means easy to answer. As soon as we attempt to do so we run up against a problem similar to the one we had when we attempted to define light. Even with the most powerful optical microscope we cannot see atoms, much less observe their structure; and so, much as in the case of light, we must resort to metaphors that 'model' the way atoms behave. There are many possible models of atomic structure, some of them very complicated. However, the behaviour we are concerned with can be described in terms of a comparatively simple model, called the Bohr atom after the physicist Niels Bohr, who first described it.

Energy states

Let us first, however, consider an even simpler model: an ordinary mousetrap. Now, a mousetrap can at any time be in just one of two possible positions: set or sprung. We may think of these two conditions as *energy states*. When we set the trap we have to put energy into the spring, and this energy is stored in the spring. A 'set' trap is not stable: the slightest touch will spring it. It may even go off by itself. When it does so, it immediately loses all of its stored energy. The energy does not vanish, of course: energy cannot be destroyed. For example, some of it becomes converted into sound energy, the sharp crack of the spring striking the baseplate. The sprung trap is in a stable state, and cannot lose any further energy.

In Bohr's model, an atom is considered as being made up of a central nucleus with a number of electrons orbiting it in fixed 'shells' (sometimes called orbitals). The maximum number of electrons that can occupy a given shell is also fixed. The shells all correspond to different energy states, and when the atom is in its normal state the electrons all occupy the shells which represent the least energy. In this condition the atom resembles our sprung mousetrap: it is stable, and unable to lose energy. We say that the atom, and its electrons, are in the *ground state*.

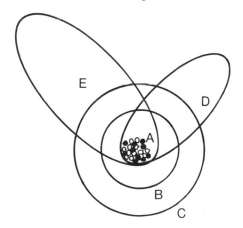

Bohr model of atom of neon (amplified). **A** Nucleus of 10 protons and 12 neutrons. **B** First electron 'shell' containing 2 electrons in ground state. **C** Second electron 'shell' containing 8 electrons in ground state. **D** and **E** Shells corresponding to higher energy states.

If an atom in the ground state receives energy by some means (such as heat, or an electrical discharge), the energy will be passed to its electrons. If the amount of energy is sufficient, one

21

or more electrons may leave the atom altogether, and the atom is said to be ionized. A gas containing large numbers of ionized atoms and free electrons conducts electricity readily; this property is made use of in gas lasers, as well as in the more familiar neon tubes and sodium or mercury vapour street lamps.

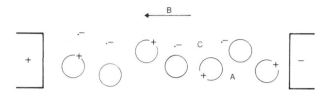

Ionization of atoms in a gas. **A** Electrons removed from atoms by strong electrical field, leaving positively charged ions. **B** Electrons have a negative charge and travel down the tube from the negative to the positive electrode, carrying a current.

If the energy is insufficient to cause ionization, it may nevertheless be sufficient to flick an electron into another shell of higher energy. The electron (and the atom) is now said to be in an *excited state*. This state is not stable, and almost immediately the electron snaps back spontaneously into its original shell, giving up its energy in the form of a photon. Of course, not all the energy is

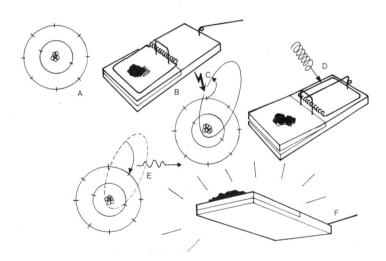

Absorption and spontaneous emission of a photon. **A** Atom of neon in ground state is analogous to a 'sprung' mousetrap (**B**). **C** A pulse of energy flicks electron into higher energy state, modelled by 'set' mousetrap (**D**). **E** Electron snaps back into ground state, releasing the stored energy in the form of a photon, just as the mousetrap may 'spring' spontaneously and release its stored energy.

converted into photons. A filament lamp consuming 100 watts (W) of electrical power emits only about 3 W of luminous power, the remainder appearing chiefly in the form of heat. However, some atomic 'events' do result in the production of photons of visible light. It is these events that concern us. The frequency of a photon is directly proportional to its energy; when an electron snaps from a higher to a lower energy state, its energy loss is the same as the energy of the emitted photon, and this determines its frequency.

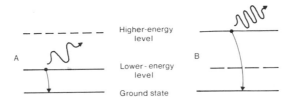

Frequency of an emitted photon. **A** An electron which snaps from a low-energy level to the ground state emits a photon of lower frequency than **B,** the photon emitted by an electron snapping down from a higher energy level. The frequency of a photon is directly proportional to its energy.

Line spectra

For most chemical elements there are a great many different electron transitions possible, and each one produces a photon of a specific wavelength. This accounts for the line spectrum emitted by an ionized gas such as neon. Each line corresponds to one possible transition; the more complicated the atom the larger the number of lines. Thus hydrogen, with only one electron, has only five lines in the visible spectrum; whereas xenon, with 54 electrons, has so many thousands of possible transitions that its spectrum is to all intents and purposes continuous. The gas used in electronic flashtubes is mainly xenon, and its discharge spectrum has a quality closely resembling that of daylight.

Continuous spectra

Incandescent materials, such as the tungsten filament of a photographic floodlamp, produce photons of all energies, and therefore give a truly continuous spectrum. This is because, in addition to light, the electron transitions produce other forms of energy, such as heat in the form of atomic vibrations. Thus the photons emitted at a transition may not carry all the transition energy, and thus cover a range of frequencies. The distribution of energy throughout the spectrum is related to the absolute temperature of the filament. This is why we are able to describe the colour quality of an incandescent source in terms of *colour temperature*, expressed in *kelvins*.

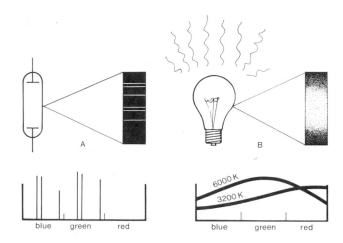

Line and continuous spectra.   **A**   Gas discharge lamp produces spectrum of discrete lines, each corresponding to a particular energy-level jump.   **B**   Incandescent lamp dissipates most of its energy as heat, the remainder being emitted as photons of a continuous spectrum of visible frequencies. The energy distribution depends on the absolute temperature of the source.

The problem of obtaining coherence

In order to produce a beam of coherent light, there are several requirements. First, we must restrict the types of electron transition to just one, so that light of only one wavelength is emitted. Secondly, the beam of light must be well collimated (spatial coherence). Thirdly, all the photons must be in phase (temporal coherence).

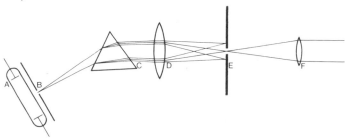

Obtaining partially coherent light with a monochromator.   **A**   Mercury vapour discharge lamp.   **B**   Small aperture.   **C**   Prism.   **D**   Lens to focus line spectrum. **E**   Pinhole allows only a narrow band of wavelength to pass.   **F**   Collimating lens.

The first two requirements can be met to some extent by a conventional gas discharge source. We can restrict the range of wavelengths by using filters, or, more efficiently, by a device called a *monochromator*. One version spreads the light into a spectrum, and a narrow slit blanks off all but a single line. Another version uses a narrow-pass optical filter to achieve the same effect.

We can obtain a fair degree of spatial coherence by using a collimating lens system. However, even the best monochromators still do not produce a beam with sufficiently high temporal coherence for practical holography. Typical coherence lengths are only about a millimetre or so, and the intensity of the beam is very low indeed. In order to produce the well-disciplined beam which we need, we have to find some method of controlling the transitions themselves.

Stimulated emission of photons

One clue to the solution had been provided as long ago as 1917, by none other than Albert Einstein (who was responsible for the original concept of the photon). So far we have considered only spontaneous emission of photons. Einstein showed that when an electron snapped from an excited state to its ground state, emitting a photon, then if that photon hit another atom in a similarly excited state, the excited electron in this second atom would also snap to the ground state, emitting an identical photon. What is more, the second photon would travel in the same direction as the stimulating photon, and in phase with it. He called this phenomenon *stimulated emission*.

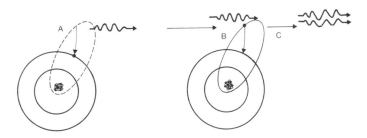

Stimulated emission of a photon. **A** Photon spontaneously emitted. **B** Emitted photon strikes another excited atom. **C** This stimulates second excited electron to snap back into ground state, emitting a photon of same frequency and phase as the stimulating photon.

This kind of event is precisely what we are looking for; for if we have enough atoms all in the correct energy state we can stimulate the emission of a photon from each, and so build up a beam of photons all of the same wavelength and in phase – a beam of coherent light. The question is, how are we to manage it?

Population inversion

The answer is *population inversion*. To see what this means, let us consider our model of the atom again. As a general rule an atom is in its ground state. Even in a gas discharge lamp, most of the atoms at any one time are still in the ground state. In order to produce stimulated emission there must be more atoms in the excited state than in the ground state – a population inversion. A photon from an atom will then have a better-than-even chance of meeting an excited atom rather than a ground-state atom (which would absorb it). Most elements have patterns of energy-level distributions which are unsuitable for producing population inversions. Fortunately, though, a few, such as chromium and neon, and one or two compounds such as carbon dioxide, do have suitable distributions; it is these substances that we use for producing coherent radiation.

The ruby laser

The first laser to be operated successfully used chromium atoms dispersed in a crystal of aluminium oxide. The crystal is, strictly speaking, an artificial ruby, though the low proportion of chromium gives it a somewhat anaemic appearance. Chromium has a transition between one of its excited states and its ground state which produces a photon of deep red light of wavelength 694 nanometres (a nanometre, abbreviated nm, is 10^{-9} metre). The chromium atom also has an energy 'band' somewhat higher than this excited state. This band consists of a large number of energy states very close together. The importance of this is that whereas the amount of energy required to raise an electron to the higher energy state needs to be exactly right, the amount required to raise it to the energy band is much less determinate, and we can use an electronic flash to provide this energy. The flashtube is usually in the form of a helix, or cylindrical spiral, with the ruby rod inside it. The energy produced by the flash is sufficient to flick nearly all the chromium atoms into the energy band; very few remaining in the ground state. The excited atoms rapidly lose a small amount of energy (as atomic vibrations) and fall to the excited energy level. We now have all the requirements for stimulated emission, and this begins at once, rapidly building up in a chain reaction.

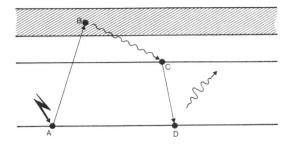

Stimulated emission in a three-level system. **A** Electron in ground state receives burst of energy, flicking it up to **B,** energy band. It loses energy (as atomic vibrations) and falls to **C,** excited energy state. From here it snaps back to **D,** the ground state, emitting a photon of appropriate frequency.

Lengthening the path

We still need a little more, however. As the ruby rod is only a few centimetres long, there will be too few photon encounters during a single transit to produce a worthwhile pulse of coherent light. The answer is to coat the ends of the rod to make them into mirrors, forming what is called an *optical cavity*. The photons are reflected back and forth between the mirror ends, greatly increasing the effective length of the rod and multiplying many times the chances of picking up further photons. One of the mirrors is made 'leaky' (50 per cent reflectance), and at each transit some of the photons pass through it, giving a beam of highly coherent light. The action ceases as soon as there are more atoms in the ground state than the excited state. A further advantage of this optical

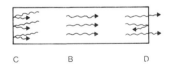

Action of a ruby laser. **A** Ruby rod surrounded by flashtube. The flash provides energy to produce population inversion. **B** Photons produced by stimulated emission travel through rod in phase. **C** Fully silvered mirror reflects all photons. **D** Partly silvered mirror permits escape of some photons.

cavity is that it acts like a resonator. Only photons with a wavelength that is an integral *submultiple* of the cavity length can take part in the amplification process. The light emitted is thus *quasi-monochromatic*.

It sounds a lengthy process, but in fact the whole operation is over in little more than a millisecond (1/1000 second). The ruby

27

laser is a *pulsed laser*; that is, it produces a brief, high-energy pulse of coherent light. If we could continue to supply the 'pumping' energy for an indefinite period we could in theory obtain a continuous beam. Without elaborate cooling arrangements, however, the crystal would rapidly overheat and be destroyed.

The helium—neon laser

The ruby laser, in common with most solid-state lasers, operates on a 'three-level' system (the ground state, the excited state and the energy band). In order to produce a continuous beam without overheating it is necessary to find some way of obtaining a population inversion without having to use so much energy. We can achieve this by operating a 'four-level' system. If we can find an atom that has a suitable electron transition between two excited states (instead of between one excited state and the ground state), we can maintain a continuous population inversion without having to use large amounts of pumping energy. This is because the lower of the two excited states will normally be empty. What happens in this case is that the atoms are flicked into the excited energy band, from which they fall to the higher of the two energy levels. They then snap down by stimulated emission to the lower energy level (laser action); and from there they quickly return to the ground state, leaving the lower energy level empty once more.

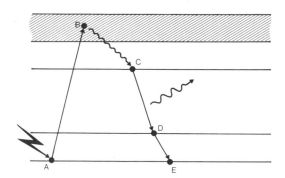

Stimulated emission in a four-level system. **A** Electron in ground state receives energy, flicking it up to energy band **B.** It loses energy (as atomic vibrations) and falls to energy level **C.** From here it snaps down to energy level **D,** emitting a photon before falling to ground state **E,** leaving level **D** empty once more.

Lasers which operate continuously are called *continuous-wave lasers* (usually abbreviated to CW), and most of them operate on the four-level principle. The most familiar one uses the gas neon as the 'lasing' element, mixed with helium at about 1/300 atmospheric pressure. The helium atoms are ionized by a continuous electrical discharge, and pass energy continuously to the neon atoms. The wavelength of the light emitted is 633 nm, in the middle of the red region of the visible spectrum.

Laser mirrors

One or both of the laser mirrors are made with a slightly concave reflecting surface, so that the beam of photons quickly becomes accurately aligned, and remains so during its many journeys through the optical cavity. The mirrors are not silvered but optically coated, with up to fifteen layers of alternately high and low refractive index, the thickness of the layers being matched to the emitted wavelength in such a way that the reflected waves interfere constructively (page 33). Such mirrors are much more efficient than metallized ones, and they have the further advantage that they reflect selectively only a narrow band of wavelengths, and over a narrow range of angles. Light of other wavelengths, or coming from other than the axial direction, simply passes through. As with the ruby laser, one of the mirrors is of 50 per cent reflectance, and it is through this that the beam of coherent light emerges. The mirrors have to be very accurately aligned, and at one time it was customary for them to be mounted in flanges equipped with setscrews. However, advances in production techniques have made it possible to mass-produce small laser tubes with permanently fixed mirrors that need no adjustment, and most present-day lasers with outputs of 5 milliwatts (mW) or less have fixed mirrors. Separate mirrors are still usual with higher-power lasers, where certain experiments may require the insertion of components into the optical cavity.

Polarization

All electromagnetic radiation can be *polarized*. The most common mode is linear polarization, where the wave vibrates in one plane only. Radio signals are linearly polarized: the plane of polarization determines whether we have to set our receiving aerials vertically or horizontally. Light can be polarized, too. Polarized light is not visibly different from unpolarized light, but it behaves differently when it is reflected from a polished non-metallic surface such as glass. When unpolarized light strikes a glass surface obliquely, part of the beam is transmitted and part reflected; both portions are partly polarized. The reflected beam

Linear polarization of light. **A** Unpolarized light waves vibrate in all planes. **B** Linearly polarized waves vibrate in one plane only.

is polarized mainly in the same plane as the surface, the transmitted beam mainly perpendicular to it. At a certain angle, called the *Brewster angle* (after Sir David Brewster, who first described the phenomenon), the polarization is a maximum for the transmitted beam and is total for the reflected beam. Consequently, if a beam of light which is already polarized perpendicular to a surface strikes it at the Brewster angle, none of it will be reflected, and it will be totally transmitted, except for the small proportion absorbed by the glass. For many purposes, including holography, it is desirable that coherent light should be polarized; in lasers this is achieved by the use of what are called Brewster-angle windows. As their name suggests, the light beam strikes them at the Brewster angle (approximately 55° angle of incidence for glass or quartz). At each passage the transmitted beam becomes more completely polarized, so that by the time the beam emerges it is effectively 100 per cent polarized.

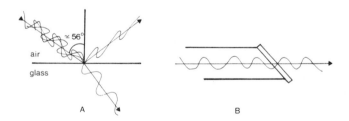

Brewster-angle reflection and transmission. When unpolarized light strikes a surface at the Brewster angle, the reflected beam is totally polarized parallel to the surface. **B** A Brewster-angle window has 100% transmittance to light polarized perpendicular to it.

Small lasers without Brewster-angle windows emit light that is randomly polarized. This is not quite the same thing as being unpolarized: the beam is in fact polarized, but the direction of polarization drifts in a random manner. This does not exclude its use for amateur holography, though there are some disadvantages, as we shall see. A randomly-polarized laser is cheaper than

30

a linearly polarized one, and gives a slightly higher output power for a given input power.

Other types of laser

There are many other types of laser. Argon-ion lasers can be made to produce two wavelengths (green and blue), and at powers of several watts. Krypton lasers can produce three or more wavelengths. Nitrogen lasers emit coherent radiation in the ultraviolet. Using carbon dioxide, it is possible to produce either pulses or continuous infrared radiation of very high power. There are giant-pulse solid-state lasers dissipating megawatts of power in a few nanoseconds, 'tunable' dye models with continuously variable wavelength control, and semiconductor lasers which can be as small as a pinhead. Such lasers are used in industry, and for scientific research; some are used for specialized types of holography. But it is the low-power helium–neon CW laser that is the most suitable for amateur work; and, fortunately, it is by a long way the cheapest, as well as the most reliable, of the lasers suitable for holography.

What is a hologram?

Wavefronts

On pages 16–19 we explained that electromagnetic radiation could be called 'coherent' if, and only if, all the photons making up the beam were in phase. The term 'phase' is a comparative one: the formal definition of phase is the relationship between the position of a wavecrest and a given reference point. If you take a beam of coherent light and join up all the points where the photons are in the same phase, you will generate a surface (known as the locus of the points) which we call a *wavefront*. In a collimated beam the wavefronts are planes: for a beam diverging from a point source they are parts of the surface of spheres. However, a beam of light which has passed through a transparent object such as a glass animal, or has been reflected from an opaque object such as a chessman, has wavefronts which are exceedingly complicated surfaces.

Diffraction

In school physics it is customary to treat the transmission and the reflection of light as separate phenomena (refraction and reflection). In the physics of waves, however, they are recognized as being basically the same phenomenon, called *diffraction*. If an object interrupts a beam of coherent light, the shape of the diffracted wavefronts is unique to that object. If part of the wavefront is intercepted by our eye, the effect is that we 'see' the object. From a different position our eye intercepts a different

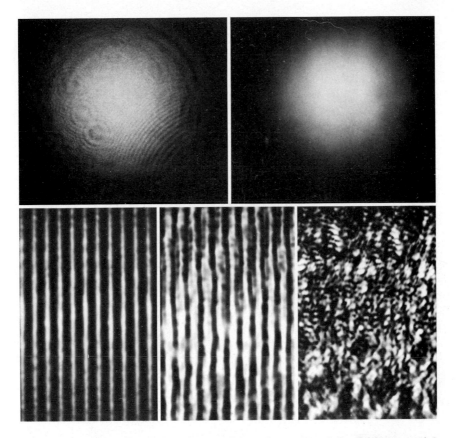

Above   The patch of light from a spread laser beam. *Top, left*   Without spatial filtering. *Top, right*   With spatial filtering. Note how the addition of a spatial filter causes the swirls and fringes to disappear. *Michael Wenyon.*

Below   Interference pattern (magnified) produced by two coherent beams overlapping at a small angle. *Bottom, left*   No modulation of either beam. *Bottom, centre*   One beam slightly modulated by a piece of fuse wire. *Bottom, right*   One beam heavily modulated by an irregular lump of acrylic material. The pattern encodes information about both the amplitude and phase of the wavefronts from the object. If this pattern is recorded on a photographic film, the information can be retrieved by illuminating the developed film with the original unmodulated 'reference' beam. *Graham Saxby.*

The sandbox built by the author. It measures 1·5 × 2 m and is shown set up for an interferometric test of stability. The piping in the foreground is part of a safety rail and is not touched by the box.

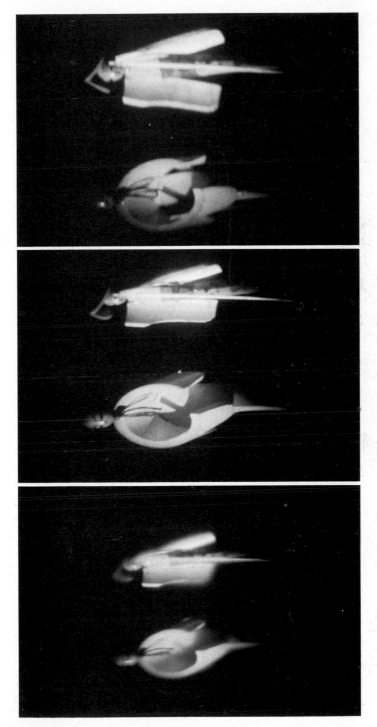

These photographs illustrate the three-dimensional quality of a hologram when viewed from different angles. The laser transmission hologram was made by Kaveh Bazargan. *Photographs by Paul Smith.*

A commercially produced white-light reflection hologram of a sectioned nautilus shell, made by the Denisyuk technique. These holograms are made using dichromated gelatin instead of silver halide, and the transparency and grainlessness of the material result in an extremely bright and almost achromatic image. Dichromated gelatin is over 1000 times less sensitive than conventional photographic emulsions, and a very powerful laser is needed for adequate exposure.

 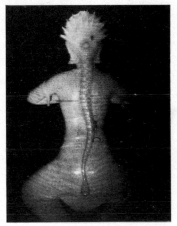

Above A white-light transmission hologram made by Kaveh Bazargan. This hologram is almost achromatic, and has the unusually large reconstruction beam angle of 80°.

Below A novel method of making a 360° hologram. A hologram of the front of the figure is exposed from one side of the emulsion, and a second hologram, of the rear of the figure, is exposed from the other side. When the hologram is set up with a reconstruction light each side, both front and rear of the figure can be seen in turn by walking round it. Hologram by P. Hariharan. *Photograph by courtesy of P. Hariharan, CSIRO, Sydney, Australia.*

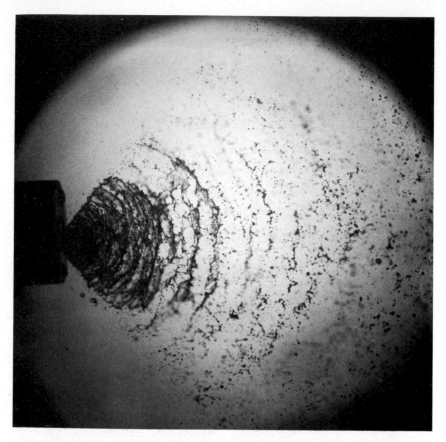

Lasers with a pulse duration of less than 0·1 microsecond can produce holograms which 'freeze' the motion of particles, such as these droplets emerging from a rocket fuel injector needle. The advantage over conventional high-speed photography is that the virtual image can be photographed with the copy camera focused on any plane within the recorded volume, enabling any individual group of droplets to be examined. By changing the viewpoint between successive exposures, stereoscopic pairs of photographs can be produced. *Photograph by courtesy of Ralph Wuerker, TRW Systems Inc.*

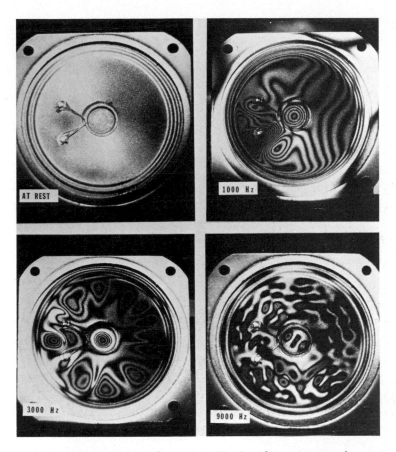

Time-average holographic interferometry. The interference pattern is generated by the images at the extremes of the speaker cone movement, where the cone is momentarily stationary. The interference fringe pattern indicates the modes of vibration at different frequencies, and these can be matched against frequency response and harmonic distortion plots, to help in improving speaker design. Similar methods are used for examining the vibration patterns of items as diverse as turbine blades and car silencers. *Photograph by courtesy of Ralph Wuerker, TRW Systems Inc.*

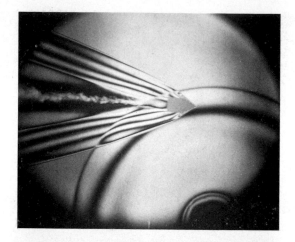

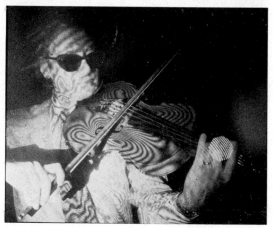

Two ways of using double-exposure holographic interferometry.
Above By making an exposure of an empty air space, then a second exposure during a disturbance of the air, the pattern of density variation is recorded as a three-dimensional interferogram. In this case it shows the interaction of two shock fronts, one from an electric spark and the other from a bullet. The event can be examined through a wide field of view. *Photograph by courtesy of Ralph Wuerker, TRW Systems Inc.*
Below By firing two pulses in rapid succession, the patterns of vibration of a musical instrument in the hands of a musician can be recorded. This hologram is one of a series made to illustrate the vibrational modes of a viola at different pitches and with different bowing techniques. The dark glasses are needed because of the very high power of the laser used here. *Photograph by courtesy of Keith Hodgkinson of The Open University and John Cookson of Loughborough University.*

part of the wavefront; as we move our eye the appearance of the object changes smoothly. This phenomenon is called *parallax*, and it is responsible for the three-dimensional appearance of solid objects. Even without changing our position we can perceive depth, for our two eyes intercept different portions of the wavefront to produce what is called stereoscopic vision. This is exploited in photography in what are called *stereograms*, which present the eyes of the viewer with discrete pictures taken from viewpoints corresponding to the two eyes.

Interference

Wherever two coherent wavefronts are superimposed, an *interference pattern* is created. At each point the intensity will represent the sum of the two waves at that point. Thus where the waves are in phase (i.e. peaks coincide with peaks, and troughs with troughs), there will be a maximum of intensity. This is known as constructive interference. Where the waves arrive in antiphase (i.e. peaks from one source coincide with troughs from the other and vice versa), there will be a minimum of intensity: this is known as destructive interference. The light and dark regions of the pattern thus correspond to in-phase and antiphase states of the two wavefronts respectively. In between these maxima and minima there will be a gradual change in intensity as the relative phases of the two wavefronts change across the screen.

It is possible to show that when two plane waves interfere in this manner, the amplitude of their resultant varies across the screen in a *cosinusoidal* manner. If the peak amplitudes of the

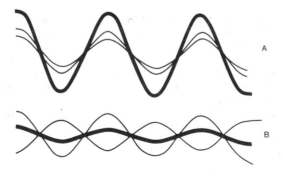

Constructive and destructive interference. **A** Two cosine waves in phase give a resultant of the same phase and of an amplitude which is the sum of their amplitudes. **B** Two waves in antiphase give a resultant which is their difference. If their amplitudes are equal the effect is total cancellation.

33

perturb = disturb, agitate, confuse.

Two waves in different phases. Two cosine waves in different phases give a resultant shifted in phase, and of a different amplitude from both. If one of the component waves is always of the same amplitude and phase (as in the above examples) the amplitude and phase of the resultant can tell us the amplitude and phase of the second, varying wave.

two beams are the same, the resulting intensity pattern will also be cosinusoidal; and by putting a piece of film in the composite beam and giving it a suitable exposure, you can produce what is called a *cosine grating*. On page 65 there is a description of how to do this, and some of the uses such a grating can have.

Encoding phase and amplitude

If you were to interrupt one of the beams with a simple object such as a flat acrylic ruler or a sewing needle, and were to examine the interference pattern with a magnifying glass, you would see that the cosine pattern had been slightly perturbed. This is because the interposed object has diffracted the beam, so that its wavefronts are no longer simple planes. A more complicated object such as a glass animal will cause much greater perturbation, so that the pattern is more violently disturbed.

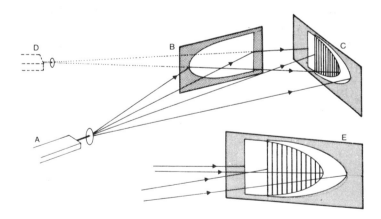

A cosine interference pattern. **A** Laser with beam spreading lens. **B** Float glass 'mirror' at a very small angle to beam. **C** Screen showing cosine pattern. **D** 'Virtual source' of second, interfering beam. **E** Screen angled to expand pattern makes it easier to see.

34

Black-and-white Plate 1 shows what happens to the pattern as a simple object (a sewing needle) and a complicated one (an acrylic figure) are inserted in the beam.

If you want to demonstrate this effect for yourself, you can produce your two 'sources' by setting up a mirror at a very small angle to a laser beam that has been slightly spread by a concave lens so that the two halves of the beam overlap. The reflected beam seems to have originated from a second, 'virtual' source alongside the first. Now, because the wavefront of the unperturbed beam (the *reference beam*) has not changed, the alteration in the pattern must tell us something about the new wave-front from the perturbed beam (the *object beam*). It is not difficult to see that any change in intensity of the fringe pattern at a given point represents a change in amplitude at that point; and any change in position of a fringe at the same point from its 'undisturbed' position indicates a change in phase of the object wave at that point. So both amplitude and phase of the object wave at the plane of the screen are encoded in the pattern, and this is all the information needed to describe the object wavefront in full. If we place a photographic film in this plane it will record this information. And if we can record it we can recreate it, as we shall see. We will have produced a *hologram*.

An analogy in sound

As this idea may be a little difficult to appreciate, let us consider an analogy. There is a form of binaural sound recording called 'dummy-head' recording. In this technique, a full-sized model of a human head, with microphones at the entrances to its ears, is positioned in the auditorium. The sound picked up by each microphone is recorded independently on two-track magnetic tape. When this is played back to a listener through headphones, the soundwaves that are recreated at the listener's ears correspond exactly with those which were picked up by each microphone and 'frozen' into the tape. The listener receives precisely the soundwaves which would have reached each ear, had he or she been sitting in the same position in the auditorium as the dummy head. The two sound wavefronts have been encoded with all the tiny differences in amplitude and phase due to their two differing positions. The two tracks thus contain all the information that tells the listener where each element of the sound is coming from. The realism, compared to even the best multi-speaker stereo sound, is startling.

Holographic reconstruction

Just as a binaural recording is a coded record of the complex sound wavefront produced by an orchestra at a certain position in space, so is a hologram a coded record of the complex light wavefront produced by an object at a certain position in space. Reconstruction from a binaural recording requires an amplifier and headphones, but reconstruction from a hologram is a much simpler matter. All that is necessary is to return the hologram to its original position, with the unmodulated reference beam, now called the *reconstruction beam*, directed at it from the original angle. The complex pattern stored in the hologram produces its own diffraction pattern; it turns out that this pattern is a precise replica of the complex wavefront that originally came from the object and illuminated the film. So when this beam reaches our eyes, each eye sees exactly what it would have seen had the original object been creating the wavefront.

If you know a little mathematics and feel you have been short-changed over this explanation, you will find a formal proof in Appendix 1.

If you shift your position, your eyes will intercept a different portion of the wavefront. You now see the 'object' as if from that new position. Thus as you move your head you see a changing view of the image, exactly as if you were looking at the original object from different angles.

Opaque objects

So far we have dealt only with a transparent object. We can equally well use the light reflected from an opaque object as our object beam, as reflected light is diffracted according to optical principles exactly the same as those which govern the diffraction of transmitted light. By causing this reflected object beam to interfere with a reference beam in the same way as before, and allowing the superposed beams to fall on a photographic film, a hologram of an opaque object can be produced; and we can reconstruct the image in exactly the same way. On looking through the hologram we will see what appears to be the original object, with full parallax.

Virtual and real images

The image is called a *virtual image*, as the light does not actually pass through the image space, but only appears to have originated from it. Holograms can also be illuminated in such a way as to produce a *real image*. The light actually passes through the space occupied by a real image, and it can be seen on a screen placed in that space. It is comparatively easy to see a virtual image, but sometimes a little difficult to see a real image, as in general a hologram produces a real image which is in front of it. It is not always easy, without practice, to focus one's eyes on the empty space in front of something. Also, the real image has some odd qualities, as we shall see.

Optical layouts for holography

The rest of this chapter is devoted to descriptions of optical arrangements for making the various types of hologram. All holograms, whatever their type, need a reference beam of coherent light as well as an object beam derived from the same source; it is the relationship between these two beams which gives us the different types of hologram.

Gabor's original layout

Gabor's experimental arrangement shows how hc managed to solve the problem of the very low coherence length of the sources then available to him. His object was a tiny circular transparency 1.5 mm in diameter bearing the names of Huyghens, Young and Fresnel. The light diffracted by the lettering formed the object beam, and the light which passed through the transparent parts formed the reference beam. The photographic plate was nearly 60 cm (24 in) away from the object, because of the very small angle between the reference beam and any part of the object beam, the optical path differences between the two was within the coherence length. When the original light source was used to illuminate the developed interference pattern a virtual image

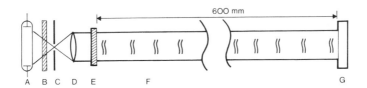

Gabor's original holography arrangement. **A** Mercury vapour lamp. **B** Monochromatic filter. **C** Pinhole. **D** Collimating lens. **E** Transparency. **F** Beam diffracted by transparency, collinear with reference beam. **G** Photographic plate.

37

appeared in the position of the original transparency. Gabor's arrangement is simple, and does not require ultrafine-grain film. Unfortunately, the view of the virtual image is hampered by the presence of a real image on the viewer's side of the hologram. Because all the optical components are in a straight line, it is impossible to get rid of the effects of this image. Nevertheless,

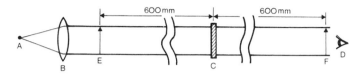

Virtual and real images in Gabor's hologram. **A** Monochromatic source. **B** Collimating lens. **C** Hologram. **D** Eye. **E** Virtual image. **F** Real image.

Gabor holography has a number of scientific applications, e.g. in particle counting.

Offset-beam transmission holograms

The first attempt to get rid of the obtrusive real image, or at least to move it out of the way, was made by Emmett Leith and Juris Upatnieks. In their arrangement the reference beam passed by the transparency to one side, and was deflected by a prism so that it overlapped the object beam. This small offset angle shifted the

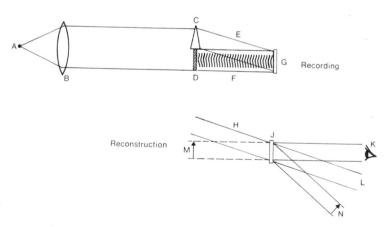

Offset transmission holography (Leith and Upatnieks). **A** Source. **B** Collimating lens. **C** Prism. **D** Transparency. **E** Reference beam. **F** Object beam. **G** Plate. **H** Reconstruction beam. **J** Hologram. **K** Eye. **L** Zero-order beam. **M** Virtual image. **N** Real image.

38

real image away from the line of the virtual image. This arrangement is almost identical with the one described on pages 34–35, the only difference being the use of a prism instead of a mirror.

An advantage of offsetting the reference beam is that the object-illuminating beam can be diffused before passing through the transparent object. Light is distributed all over the film area instead of in discrete patches; this results in a much better image reconstruction.

These early experiments were carried out using mercury vapour light; further improvements in technique had to await the arrival of light sources of greater coherence length. With the appearance of the laser it became possible to make holograms with a *depth of field* of several centimetres.

Reflected-beam transmission holograms

The advent of the laser also made possible holography of solid objects, the object beam being reflected from the object rather than transmitted through it. This is the arrangement we shall be using for our own transmission holograms of opaque objects.

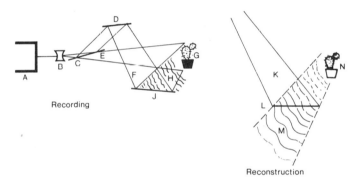

Arrangement for holography of opaque objects. **A** Source. **B** Beam spreader. **C** Beamsplitter. **D** Mirror. **E** Object illumination beam. **F** Reference beam. **G** Object. **H** Object beam. **J** Film. **K** Reconstruction beam. **L** Hologram. **M** Image beam. **N** Virtual image.

Using this arrangement it is also possible to make transmitted-light holograms by placing a white diffusing reflector behind the transparent object.

Large holographic studios use massive surface tables to support optical components on heavy bases; in general, they use lighting arrangements more complicated than those we have studied so far. The object illumination may consist of several beams, not necessarily in one plane, obtained from one laser beam by means

39

of *beamsplitters*. Each beam has its own expanding lens, and imaginative lighting arrangements such as those used in photographic studios are possible. There will usually be three object

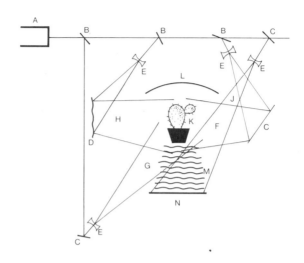

Typical holographic studio lighting arrangement. **A** Laser source. **B** Beamsplitters. **C** Mirrors. **D** Matt reflector. **E** Beam spreaders. **F** Main object illumination beam (elevated). **G** Second object illumination beam (horizontal) **H** Fill-in object illumination beam (diffuse). **J** Reference beam (elevated). **K** Object. **L** Background. **M** Object beam. **N** Holographic film or plate.

beams. One of these is elevated and plays the role of a modelling light. A second may have the role of a spotlight, while the third is bounced off a diffusing reflector to provide the fill-in light. It does not matter how many object-illuminating beams there are, provided they all have approximately the same optical path length from the laser to the film as the reference beam.

White-light reflection holograms

In Gabor's original in-line holography the angle between the object and reference beams was only a small fraction of a degree; and in Leith's and Upatnieks's arrangement the angle was only a few degrees. The interference patterns generated were coarse, and it was not necessary to use particularly fine-grain film. But in reflected-light holography the angle between the beams may be 60° or more, which means that the fringe separation is roughly the same as the wavelength of light. Modern holographic emulsions have a mean particle diameter (unexposed) of about 30 nm (3×10^{-8} m) which is about one-twentieth of the wavelength of

40

light. This leaves plenty of resolution capability to spare, even when we take into account the clumping of the grains during processing.

If we make the reference beam enter the emulsion from the opposite side to the object beam, we get an entirely different type of hologram. The fringes will now be only half a wavelength apart, will be distributed throughout the thickness of the emulsion, and will be parallel to its surface instead of perpendicular to

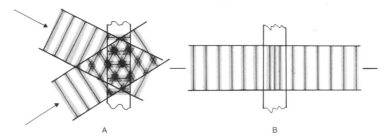

Orientation of fringes in transmission and reflection holograms. **A** In a transmission hologram the fringes are roughly perpendicular to the emulsion surface and are about one wavelength apart. **B** In a reflection hologram they are parallel to the surface and half a wavelength apart.

it. A reflection hologram requires a different optical arrangement from a transmission hologram, as the object and reference beams illuminate the film from opposite directions.

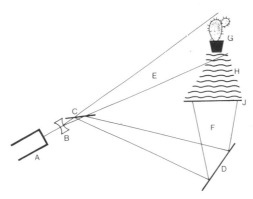

Optical arrangement for reflection holography (schematic). **A** Laser. **B** Beam spreader. **C** Beamsplitter. **D** Mirror (elevated). **E** Object illumination beam. **F** Reference beam. **G** Object. **H** Object beam. **J** Film.

This type of hologram is sometimes known as a Lippmann hologram, after Gabriel Lippmann, who invented a system of colour photography based on interference layers of the kind

41

produced here. Lippmann discovered that the regular layers of silver formed a kind of mirror that was highly selective, reflecting only a narrow band of wavelengths. For example, fringes 250 nm apart will reflect only light of a narrow band of wavelengths

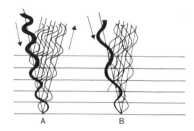

The reflection hologram as an interference mirror. **A** Light waves of the correct wavelength (twice the fringe spacing) interfere constructively when reflected by the fringes, and reinforce. **B** Light waves of other wavelengths interfere destructively and cancel.

centred on 500 nm. The image forms an *interference mirror*, similar in principle to the mirror used in lasers (page 29). Light reflected from the successive surfaces interfers constructively if, and only if, it has the appropriate wavelength. This means that we can use white light for our reconstruction beam, provided it is fairly coherent spatially. The light from a slide projector, or the sun, or even a torch bulb, will do. The hologram selects only the required wavelength, rejecting the others and giving the temporal coherence we need for the reconstruction. The hue may be yellow or green rather than red, owing to shrinkage of the emulsion

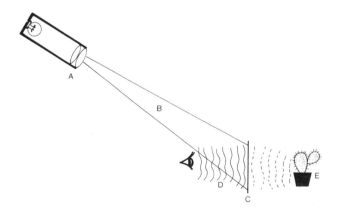

Viewing a reflection hologram. **A** White light source. **B** Reconstruction beam. **C** Hologram. **D** Image beam. **E** Virtual Image.

42

during processing; but it will still be nearly monochromatic, and the image will be sharp.

Pseudoscopic real images

If you view a reflection hologram reversed, that is with the reconstruction beam coming from the opposite direction, you will be able to see the real image in front of the hologram. But there is something very odd about it. It has reversed parallax: as you move your head the front part of the image appears to move relative to the rear, but in the wrong direction. In addition, you have to focus your eyes closer to get the back of the image sharp, not farther away as you might expect. The parallax is reversed: what we have is a *pseudoscopic image*.

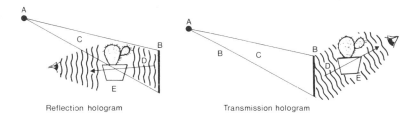

Reflection hologram   Transmission hologram

Viewing the real image.   **A**   Light source (monochromatic for transmission holograms). **B**   Hologram reversed (i.e. back turned towards reconstruction beam). **C**   Reconstruction beam.   **D**   Image beam.   **E**   Pseudoscopic real image.

Transmission holograms can also be made to produce a pseudoscopic image, and you can see it in the same way, by turning the hologram round in the reconstruction beam. The reference and reconstruction beams need not be collimated, provided the angle of spread is small and the object close to the film. However, if the hologram was made with a strongly diverging reference beam, you may need to use a converging reconstruction beam in order to see the real image clearly.

You can capture this real image by putting a photographic film in its plane. As the image is three-dimensional, it is possible to focus on various parts of it by changing the plane of the film. You can also take a photograph of the real image using a camera, just as you can of the virtual image.

Producing an orthoscopic real image

The extraordinarily lifelike *orthoscopic* (right way round) *images* shown in exhibitions of holography are made by producing a second hologram from the first, using the pseudoscopic real image of the first hologram as the object for the second. The real image given by the latter, having been reversed twice, will now be orthoscopic. By imposing certain restrictions on the object beam we can produce a transmission hologram that can be viewed by white light. The principles are discussed in the practical section (pages 98–105).

Single-beam reflection holograms

In 1962 the Soviet scientist Yuri Denisyuk discovered a method of making reflection holograms that needed no beamsplitter or mirror. The earliest results, made with only partially coherent light, were not very impressive; but with modern methods and materials it is possible to produce holograms of outstanding quality. In this configuration a single beam fulfils the role of both object and reference beams. The reference beam passes right through the emulsion to become the object-illuminating beam on the other side. Thus we have the conditions for a reflection hologram, produced in a simple and elegant way. Because of the difference in optical path between the object and reference beams the depth of field is restricted, but because the object is situated very close to the film a single-beam reflection hologram offers a large amount of parallax on a comparatively small film

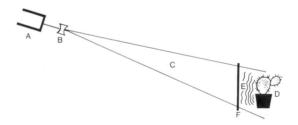

Arrangement for making single-beam reflection holograms. **A** Laser. **B** Beam spreader. **C** Reference beam. **D** Object. **E** Object beam. **F** Film.

area. Because of the very small separation of the fringes it is necessary to use the finest-grain emulsion available; fortunately, these emulsions are almost completely transparent.

Phase holograms

Single-beam reflection holograms are easy to produce, and very effective. However, for the best results they need to be processed as what are called *phase holograms*. Unlike transmission holograms, which can be processed satisfactorily in almost any developer, even monobath solutions, reflection holograms processed in the conventional way are dark, and unless the recon-

44

struction beam is intense it is hard to see the image. If, however, we replace the fixing process by a bleach bath which converts the developed silver grains into silver bromide or some other transparent substance, we can render the hologram fully transparent. You might think this treatment would destroy the image; but it does not. What happens is complicated, but a general explanation is as follows:

The method of action of a developer is twofold. In chemical development the solution attacks the particles of silver halide in the emulsion which bear a latent image, reducing them to grains of silver (you can think of a latent image as a tiny area of damage within a particle caused by absorption of photons). In 'physical' development the solution dissolves some of the unexposed particles (i.e. those which do not bear a latent image) and redeposits the dissolved silver on the developing grains. Both actions take place simultaneously. There is thus a change in the quantity of material present at any point, and if the exposure is correctly calculated it is possible to cause almost all the particles not bearing a latent image to disappear and to re-emerge in the form of metallic silver in the dense regions. The bleaching process, properly called rehalogenizing, converts the entire image into silver bromide, which is transparent and has a refractive index much higher than that of gelatin. If you remember the description of laser mirrors (page 27) you will recall that the 'interference' coating consists of layers of material of alternately high and low refractive index. These layers, one-quarter of a wavelength thick, produce constructive interference in the reflected beam for the wavelength concerned. The effect of the development and bleach is to produce just the right amount of silver bromide to act in this way. The exposure has to be right; but the result is a dramatic improvement in the brightness of the image.

This method of processing has evolved comparatively recently, and requires a very full exposure. An alternative method is to remove the developed image. There are several techniques. One type of bleach removes the silver image, leaving the undeveloped silver halide to act as the interference mirror; another converts the fixed image back into silver halide; a third removes the silver entirely, leaving the phase information in the form of complex cross-linkings in the gelatin which modify its thickness and refractive index in an appropriate manner. Many of the older methods of processing reduce definition and change the colour of the image, but they all increase the image brightness considerably. A selection of the most reliable formulae appears in Appendix 3.

Part 2

Making your own holograms

Single-beam transmission holograms

Buying a laser

Small lasers of 0.5 to 1.0 milliwatt (mW) power output and random polarization are quite satisfactory for small holograms up to about 5 × 6.5 cm (2 × 2½ in); though for larger holograms the required exposures for such lasers are so long that stability becomes a serious problem. These small lasers are not expensive, and are chiefly used by schools and colleges in the teaching of physics. In fact, if you work in an educational establishment which has a physics department, you may well be able to obtain access to one of these small lasers. If so, it would be an excellent idea to make your first small holograms as described in pages 53–72, using a small laser of this type, before you decide to take the plunge, build yourself a sandbox, and invest in a more powerful source. When you decide to buy a laser, you will need one of at least 5 mW output, linearly polarized, and without special brackets or supports. This is the lowest power that can be used for serious work; if you are thinking of making 20 × 25 cm (8 × 10 in) holograms you could well go up to 15 mW. Above 5 mW, however, the price of a laser increases much faster than its power; so for the occasional venture into large sizes you might be better off to hire one. Some addresses of suppliers of lasers, as well as of other useful equipment, are listed in Appendix 2. Most suppliers also sell beamsplitters and spatial-frequency filters (pages 83–86), and some offer holography kits ready-made. These are usually expensive and limited in scope, and you can make your own equipment for a fraction of the cost. It might be a good idea, though, to buy a sample transmission hologram, to give yourself an idea of what one looks like, and to practise setting up lighting suitable for viewing. By the way, you may be puzzled by the phrase 'TEM$_{00}$ mode' which appears in some lists of lasers. This refers to the pattern of the spot, and all small helium–neon lasers are of this type.

You can buy laser tubes separately, and if you have had some experience of building electronic equipment it is not at all difficult to build your own power supply from a circuit diagram or kit. This can save you up to one-third of the cost of the laser.

Safety precautions with a laser

If you are in the habit of watching television space-fiction programmes, you may be a little disappointed to find that your laser produces, not a blinding explosion of light, but a little red spot seemingly not much brighter than that provided by a pocket torch. Do not be misled by appearances: the beam from even a 0.5 mW laser has about the same intensity as the sun. So do not in any circumstances stare into the unspread laser beam. In fact, even catching a momentary glimpse of it reflected off a polished surface can be an unpleasantly eye-watering experience, and with a 5 mW beam can actually do permanent damage to your eye. If you are working with such a laser, wear laser eyeshields when you are setting up with the beam unspread. Make sure nobody can walk into the room and accidentally look into the beam, whether direct or reflected; and ensure nobody else can operate the laser in your absence. Remove the switch key, if there is one; otherwise, lock the door. Never operate any laser with the cover off. The power supply operates at several thousand volts.

Care of your laser

Lasers are fairly robust and will withstand the odd accidental bump; but a very tiny misalignment within the optical cavity will seriously affect the intensity of the beam, so treat your laser with respect. Before you use it for holography ensure the optical window is clean. A clean dry watercolour brush is best for removing dust. Clean the window occasionally using a cotton bud dipped in isopropyl alcohol or tape-head cleaning fluid, and wipe gently with lens cleaning tissue.

If your laser tube is of the resin-sealed type, and you are not using it regularly, it is advisable to switch it on for 15 min or so once a week; otherwise you may find, after a period of non-use,

50

that it does not begin to function until after several minutes have elapsed. This does not happen with modern glass-sealed tubes.

The usual life of a laser tube is between 5000 and 10 000 hr. When it eventually fails you can have it replaced, just like a television tube. The cost of replacement is about half the original cost of the laser.

Getting to know your equipment

Before you begin making larger holograms you need to become thoroughly familiar with the operation of your laser and the manipulation of the basic optical components. Making small holograms with straightforward optical arrangements will give you the experience you need.

Freedom from vibration

The results of one's first efforts in holography are not always altogether encouraging. The image may be dark, lacking in contrast, unevenly illuminated, ill-defined or even non-existent. Even a good-quality image may look grainy or speckled. There is not very much you can do about the speckles; this is a phenomenon tied up with the nature of coherent radiation, an interference pattern associated with microscopic roughnesses of the object's surface. The remaining defects, though, are avoidable. The darkness and lack of contrast are caused by incorrect exposure or viewing angle, the unevenness by dirty optical surfaces, and the blurring or absence of image by movement of the object or the optical components during exposure. This can be a serious problem if not tackled right at the start.

Large holographic laboratories employ massive concrete tables isolated from vibration by thick rubber blocks. They have cement floors and are usually situated in a basement. A kitchen table or carpenter's workbench standing on a wooden floor is a poor substitute for this, especially if there is a freezer or other source of vibration standing on the same floor, or if the house adjoins a busy main road. Even the air circulating from a convector heater can cause a good deal of trouble. You can get some idea of the importance of freedom from vibration from the fact that any disturbance of the optical paths of more than about one-eighth of the wavelength of light (about one ten-thousandth of a millimetre) will seriously affect the quality of a holographic image, and a movement of just four times this distance will wipe it out altogether.

It is therefore most important to isolate your equipment from all possible sources of vibration. A well-tried way of achieving this is to fix the optical components to a heavy chipboard plank or the smooth side of a concrete paving slab about 45 × 90 cm (18 × 36 in) in size, balanced on a partly inflated motor-scooter or kart inner tube. Your local garage will let you have an old one for a few pence, and will probably inflate it for you too. It should be firm but not hard. If you later go on to do sandbox holography you can use the same method of isolation, though because of the extra weight you may need up to six tubes.

The beamspreader

This is needed to diverge the laser beam sufficiently to illuminate the object and film. A 6 mm focal length concave lens will give the right amount of divergence. A ×40 microscope objective is even better (you will need one for the sandbox). But don't rush out and spend a lot of money on an expensive apochromat. Go for the cheapest student-quality objective, which is perfectly satisfactory for the narrow quasi-monochromatic beam of your laser. You can often pick up a second-hand objective for a few pounds, but make sure there is no mould on the lens surface, otherwise it will be pitted and useless.

Other optical requirements

You also need a piece of 3 mm or 6 mm float glass about 50 mm (2 in) square and a front surface mirror not less than 100 mm (4 in) square. You can get the glass from any glazier, and the other items from one of the addresses listed in Appendix 2. You may also need a *neutral-density filter* to cut down the intensity of

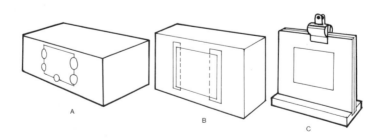

Making a sheet film holder. **A** and **B** Temporary holders using drawing pins in a wooden block or double-sided cellulose tape on glass. **C** Glass sandwich filmholder. A 6 × 9 cm sheet filmholder is also satisfactory and is easy to load.

the reference beam, and you will need a filmholder. The best way to support the film is between two glasses hinged together with tape and secured with a bulldog clip, but you may be able to get satisfactory results using a 6×9 cm ($2\frac{3}{8} \times 3\frac{1}{2}$ in) sheet film-holder, or by taping the film to a glass block.

Films and plates for holography

Both Eastman Kodak and Agfa-Gevaert produce holographic emulsions suitable for helium–neon laser light: the availability from the respective companies varies somewhat from one country to another. Appendix 3 lists the types available. Prices are somewhat higher than those of amateur films. Plates are more expensive still, and you cannot easily cut them into smaller sizes. For your early experiments use 4×5 in sheet film cut into four pieces $2 \times 2\frac{1}{2}$ in in size. The smaller the setup and the closer together your optical components, the less chance there will be of movement during the exposure; and if anything goes wrong you will not be much out of pocket.

It is not always easy to tell which is the emulsion side of a holographic film. There is a code, however, which is used for all photographic sheet films. You will find a notch near the end of one of the shorter sides of the sheet of film. If you hold the film so that this notch is on the upper edge at the right-hand corner, the emulsion side will be towards you. The emulsion side is often slightly concave, though whether this is so depends to some extent on humidity. It is a good idea to keep all your cut pieces of film in the box emulsion downwards, so that you will not be in any doubt which is the emulsion side when you come to use them. Roll film is always wound emulsion inwards. You can identify the emulsion side of a plate by tapping it between your top and bottom teeth. The glass side feels and sounds harder than the emulsion side; and if you bite the corner of the plate gently the emulsion side will stick to your teeth.

For your first experimental holograms you can use a very simple arrangement which requires no optical components other than the beamspreader, and has therefore little that can go wrong. Set up the laser and fix the 6 mm concave lens to the aperture with Blu-tack. Blu-tack (Superstuff in the United States) is a substance that superficially resembles modelling clay, but is much more adhesive. It is intended primarily for fixing posters to walls; but it is particularly good at holding optical components firmly, and we shall be mentioning it frequently. If you are using a microscope objective it will need additional support. Place your

53

filmholder in one half of the beam at a point where the diameter is about 12 cm (5 in). This is about 90 cm (36 in) from the laser. If you are using the optical bench as a support you should manage this; but if your base is too short, add a further concave lens of focal length 10–20 mm a few centimetres farther away to increase the divergence. Put some small transparent objects in the other half of the beam, close to the filmholder but not casting a shadow on it. These may be small phials of liquid or glass stoppers: small glass animals make good objects, as long as they are not dark. A small magnifying glass in front of the objects is very effective. Set the filmholder at about 20° incidence to the beam, angled towards the objects. Fix the objects to a wooden block as a plinth, using Blu-tack or a spot of 10-second adhesive.

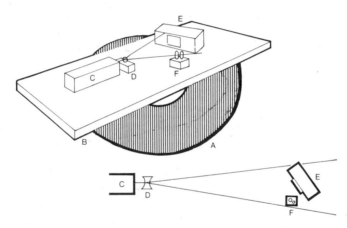

Arrangement for single-beam transmission holography. **A** Scooter inner tube acts as antivibration mounting. **B** Chipboard plank or paving stone. **C** Laser. **D** Beam spreader. **E** Sheet filmholder. **F** Transparent objects.

Put a piece of white card in the filmholder and move the object around until the reflection on the card is brightest. Avoid getting bright specular reflections on the card. When you are satisfied with the arrangement, fix the plinth in position with blobs of Blu-tack.

When making single beam transmission holograms, the intensity of the reference beam should be between three and ten times that of the object beam, the ideal being about 7:1. If you have an exposure meter, you can measure this ratio. Place the meter cell (set for reflected light, not incident light) in the position where the filmholder will go, and note the reading. Now blank off the direct beam and note the reading again. As long as there is no more than 3½ stops difference you should get a reasonably good

54

hologram. If your readings differ by much more than this, you will have to cut down the direct beam in intensity by inserting a neutral-density (ND) filter. These are grey gelatin squares which cut down the light by a fixed amount. A ND 0.3 filter has a transmittance of 50 per cent; a ND 0.6 25 per cent; and a ND 1.0 10 per cent. If your laser beam is linearly polarized you can use a piece of Polaroid as a variable filter, inserting it in the beam and rotating it until the beam is attenuated sufficently.

Making a set of test exposures

When you are satisfied with the lighting ratio, block off the laser beam with a piece of black card. Leave the laser switched on. Insert a piece of film in the holder and wait for at least 2 min, to allow the equipment to settle down. Using the black card as a shutter, give exposures to successive films of 12, 8, 5, 3, 2, 1 and $\frac{1}{2}$ sec, and a final one as quick as you can. If you are not very good at estimating seconds, try saying to yourself 'A hundred and one, a hundred and two . . .', etc (do not talk aloud, though, or the vibration will disturb the setup). You do not need a whole piece of $2 \times 2\frac{1}{2}$ in film for each exposure, of course. Cut your pieces of film horizontally in half, to make strips $1 \times 2\frac{1}{2}$ in; this way you will get eight test strips out of one 4×5 in film. Mark the exposure on the back of the film, using a wax pencil.

Processing the film

Holograms are processed in exactly the same way as ordinary black-and-white negatives. You can use any proprietary develop-er and fixer, even monobath developer, though when you come to do more exacting work you will get the best results by following the methods detailed in Appendix 3. You should process the films without delay, as the latent image fades quickly, and can disappear completely in 24 hr. Switch off the light (you can use a green safelight), unload the exposed film, and immerse it, emulsion side uppermost, in a dish of developer. Rock the dish gently for 2–3 min at about 20°C (68°F). Remove the film from the developer, rinse it briefly under the tap, and immerse it in a dish of fixer for 3–5 min. Wash the film in running water for about 10 min, then hang it up to dry, removing any drops of water with a cellulose sponge. Handle the film throughout by the edges only. If you wish you can speed up drying by using a hair dryer on the films (but do not do this when you come to make reflection holograms later).

Viewing your holograms

Choose an exposure that is a middle grey in tone, and hold it in the spread beam of the laser, as nearly as possible in the original position and at the same angle. Look through it towards the point where your objects were positioned. You should be able to see

Viewing the hologram (reconstruction). **A** Laser. **B** Beam spreader. **C** Developed holographic film set up in original position. **D** Diffracted waves from hologram reach eye, appearing to have originated from **E** Virtual image.

the image quite clearly. Try all of the exposures, and select the one which gives the brightest image. Make a note of the exposure time against your note of the meter reading. You can now use this information to help you to obtain a correct exposure for any other meter reading.

Possible causes of failure

If there seems to be no image at all, check that you have set up the film the right way round, exactly as it was in the filmholder. Change the orientation a little. Change your viewpoint. Finally, if all else fails, get somebody else to have a go. He or she may find the image you have somehow missed.

If there is still no visible image, and your films are rather dark, perhaps all your exposures were excessive. Repeat the tests with a 1.0 ND filter over the laser window. This will cut the exposure by a factor of 10. The average *density* of a hologram of this type should be in the region of 0.6–0.8. You can check this with your exposure meter as follows: First point the meter at an evenly illuminated plain wall, and note the reading. Now place the exposed film over the meter cell and note the reading again. The density will be correct if the difference between the readings is about $2\frac{1}{2}$ stops.

If one of your holograms has approximately the correct density, but nothing you do produces an image, there must have been movement during the exposure. To check this, view the hologram as you did before, but using light from the sun or a small lamp bulb. At one particular orientation you should be able to see rainbow-coloured smears. If no colours appear, no matter how you move the film around, then it does not contain any interfer-

56

ence fringes, and this means they have been wiped out by movement during the exposure. Possible causes of this include:

1 optical component support not completely isolated from vibration of floor;
2 loud sounds causing one or more components to vibrate;
3 warm air currents from a radiator or refrigerator, or your own breathing;
4 draught, from uncovering laser beam too enthusiastically;
5 laser not fully settled down: it needs 10–15 min after switching on.

See also the list of Do's and Don'ts on page 107.

Uneven illumination

You will probably have noticed a number of dark swirls and irregular blotches in the disc of illumination. These are caused by specks of dirt on the optical surfaces, and optical imperfections in the beamspreader lens. They will not prevent your obtaining holograms, though they will be visible when you view the image. Try cleaning the lens with a cotton bud dipped in isopropyl alcohol or tape-head cleaning fluid. Clean the laser window, too: be very gentle. If the swirls persist, try rotating the beamspreader until the illumination is as uniform as possible across the film frame area.

Reflected beam transmission holograms

Building your own optical bench

In university physics courses basic holography is usually taught using an optical bench. Although a chipboard plank is a reasonably satisfactory support for optical items in small-scale holography, an optical bench is better. When you come to make larger holograms with a sandbox you will need a rigid support for the laser and relay mirror, and the optical bench described below is ideal. You can build it in half an hour, using materials you can pick off the shelf at any large plumbers' and builders' merchants. The material you require is 15 mm (formerly 'half-inch') stainless steel or hardened copper tubing. You will need two lengths of about 1.1 m (7 ft). You will also need four right-angle York elbows; sixteen nylon pipe-clips and woodscrews; about 1.5 m (5 ft) of 15 × 100 mm ($\frac{5}{8}$ × 4 in) timber or floorboarding, a piece of heavy-duty chipboard about 300 mm (12 in) square, and a length of curtain spring with four hooks and four screw-eyes. You will also need an assortment of wooden blocks for holding your components, and a supply of Blu-tack.

Cut a 75 mm (3 in) length from each of the tubes to act as crosspieces for the framework. Fix the crosspieces and the long tubes into the York elbows with epoxy resin adhesive, cut the plank to length, and fix the frame to it with four of the pipe-clips. Make the laser mount from the remainder of the plank and secure it to the optical bench with four pipe-clips. Anchor the laser with two short lengths of curtain spring, using the hooks and eyelets to secure the ends. Fix the chipboard to the other end of the optical

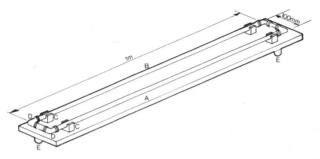

A simple home-made optical bench. **A** Base of 15 × 100 mm ($\frac{5}{8}$ × 4 in) timber. **B** 15 mm ($\frac{1}{2}$ in) stainless steel water pipe. **C** 15 mm nylon pipe clips. **D** 15 mm elbows. **E** Door stops (3).
NOTE: This bench will become the laser and relay mirror support for your sandbox.

bench with four pipe-clips. The remaining four clips are for the relay mirror for your sandbox, and you will not need them at the moment.

You will find it well worthwhile taking the small amount of trouble needed to make the optical bench. It is invaluable for

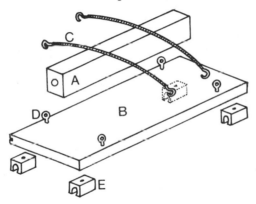

Support for laser on optical bench. **A** Laser. **B** Floorboarding cut to length. **C** Curtain spring with hooks. **D** Eyelets. **E** Nylon pipe clips.
NOTE: Throughout the diagrams in this book, the laser is shown conventionally, i.e. rectangular in cross-section. The majority of small lasers are in fact cylindrical. This does not affect the method of support.

holding the components firmly for making holograms, and it can be used for some of the fascinating optical experiments possible with coherent light, which space precludes our describing here, but which you can find described in up-to-date courses such as The Open University's 'Images and Information' course (see Appendix 5). However, if you do not feel like building one right away, you can still make small transmission holograms without it. In fact, when I was working on the early drafts of this book I

59

carried out all the preliminary experiments on an ordinary workbench, using empty grape-juice tins as supports, the components being fixed to them with large lumps of Blu-tack. But the optical bench certainly made things easier.

Method of setting up

Transmission holograms of transparent objects are fairly simple to set up and expose, and they produce a satisfactory three-dimensional effect. However, both the lighting arrangement and the choice of subject matter are limited. Reflected-beam transmission holograms offer more scope for creativity. In this type of hologram the object beam is reflected from the object rather than transmitted through it. The intensity of the object beam is quite low, and it is usually necessary to cut down the intensity of the reference beam correspondingly. We can achieve this by using an unsilvered beamsplitter, supplemented as necessary by a neutral-density filter. We also need to ensure the optical paths of the object and reference beams are equal to within a centimetre or so; in order to achieve this we will need to fold the path of the reference beam.

Until you have had some practice in setting up reflected-beam transmission holograms you may find that getting the optical paths right can be a bit tricky. As a guide, you might find it helpful to trace the diagram opposite, and use your tracing as a template to mark out your area.

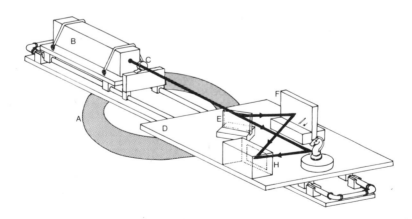

Arrangement for reflected-beam transmission holograms (shown using home-made optical bench). **A** Scooter or kart inner tube. **B** Laser. **C** Beam spreader. **D** 300 mm square support table. **E** Float glass beam splitter. **F** Front surface mirror. **G** Object on plinth. **H** Filmholder. For clarity the beam is shown unspread.

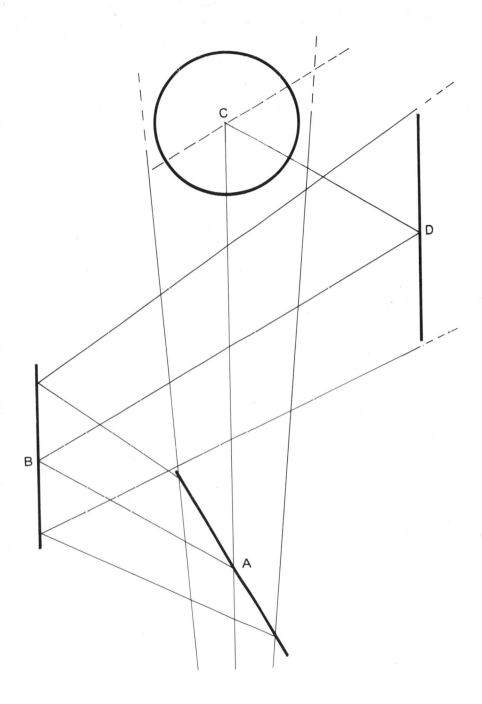

Layout for transmission holograms (actual size). **A** Beam splitter. **B** Mirror.
C Object area. **D** Filmholder. Distances **AB, CD** = 30 mm; **AC, BD** = 60 mm.
Distance from **A** to beamspreader = 140 mm approximately (for a 6 mm lens).

61

In addition to your beamspreader and filmholder you will need a beamsplitter and a front surface mirror. Use 3 mm or 1.5 mm float glass for the beamsplitter, or a 50 × 50 mm slide cover glass. Mount this and the mirror in tight-fitting saw-cut slots in wooden blocks. If you are using a 6 mm concave lens or a ×40 microscope objective as a beamspreader, the distance from the beamspreader to the centre of the beamsplitter will be approximately 14 cm (5½ in).

Align the optical components using the unspread beam. First, ensure the object-illuminating beam passes through the centre of the object space. Next, align the beamsplitter so that the beam passes through its centre and strikes the centre of the mirror (you can check this with a piece of tracing paper), and then align the mirror so that the spot falls on the centre of the filmholder. Now insert the beamspreader, and align it so that the film area is adequately covered. It does not matter seriously if the corners of the film area are dark, as long as the beam reaches to the edges.

Choose a small bright object such as a metal toy (or, if you are a traditionalist, a small white chessman!). Put it on the plinth in the correct position, ensuring it is fully illuminated and is clear of the reference beam.

If you have an exposure meter, you can check the relative intensities of the reference and object beams by blocking off each in turn. Ideally the ratio should be about 3 to 1, the reference beam being the higher intensity, but anything between 2 and 10 to 1 will give you holograms of reasonable quality. Also check the total intensity of the two beams the way you did for single-beam holography (page 54) for future reference. With this setup the intensity of the reference beam can sometimes be too high, by a factor of up to four. If so place a 0.3 ND filter close to the mirror so that the reference beam passes twice through it, then the beam will be attenuated by about the right amount. Tilt the filter so that any light reflected from it will not reach the filmholder, and fix it in place with Blu-tack.

Exposure trials

You are now ready to make your test exposures. Carry them out in the same way as described on page 55 and process the film as described also on page 55.

To examine your holograms, hold them in the spread laser beam at the same angle as the original reference beam. You may find this easier if you use the optical bench, especially if you make a holder set at the correct angle. Look through the hologram in the direction of the original objects. Change the direction of the hologram relative to the reconstruction beam until the image is at its brightest. Try all the exposures. The one which gives the brightest result is the correct exposure; and if you noted the meter reading, you will be able to estimate exposures correctly even when the illumination level is different.

Finding the real image

Once you have succeeded in getting a good quality hologram, try to find the real image. Reverse the hologram and shine the unspread beam through its base side holding a piece of ground glass or tracing paper on the side away from the laser and taking care not to look into the beam. You will be able to catch the real image on your screen. It will be to one side of the spot, looking much the same as the image formed by a conventional lens. As you move the hologram around in the beam the apparent viewpoint will vary correspondingly: if the beam goes through the top of the hologram you will see the image as if viewed from above, and if it goes through the side you will see the image as if from that side.

If you have a 150 mm or 200 mm lens (a simple lens or a camera lens) position it in the spreading beam so that it gives a beam which converges at about the same angle as the spreading beam diverges. Place your reversed hologram in this beam, close to the lens, and examine the real image. You will notice that parts of the image are now out of focus, and that you have to move the

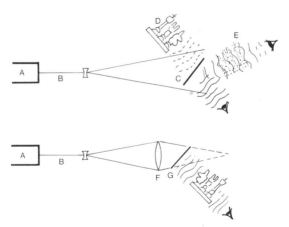

Viewing the virtual and real images. **A** Laser. **B** Beam spreader. **C** Hologram. **D** Virtual image. **E** Pseudoscopic real image (difficult to see). **F** Converging lens. **G** Hologram reversed. **H** Pseudoscopic real image (easy to see).

screen nearer to, or farther from, the hologram in order to focus different parts of the image. The odd thing is that you have to move the screen in what seems to be the wrong direction: nearer to get the farther parts in focus, and vice versa. This is the pseudoscopic effect referred to on page 43. If you now take away the paper without changing the focus of your eyes, you will be able to see the real image directly, floating in front of the hologram. It may be somewhat magnified or diminished, depending on the degree of convergence of the beam. It is possible to capture this image on a photographic film, though this is not as easy as photographing the virtual image with an ordinary camera.

Once you have managed to 'catch' and hold the real image visually, try moving your head, still watching it. Because of the reversed parallax, the image seems to take on a life of its own, apparently rotating in the same direction as you move your head, but twice as fast. It is possible to duplicate this illusion in the 'real' world (page 70).

Phase holograms

You can make your images much brighter by bleaching them to produce phase holograms (page 44). The optimum exposure for a phase hologram is greater than that for an amplitude hologram. Typical average density for correct exposure is about 2.0, requiring about three times as much exposure. You can use an exposure meter to check the density before bleaching: with the hologram over the meter cell it should read about $7\frac{1}{2}$ stops, or 100 times, less than without the hologram.

'Table-top' pictorial holography

Once you have managed to produce a good-quality hologram, and feel confident about your technique, you can begin to be creative. Of course, your object space is only about 50 mm in diameter, which limits you to studies of the 'table-top' kind. Early holographers seemed to have a fixation about chessmen; and you may well want to try some compositions with miniature chessmen, too. Small toys such as model cars or planes are fairly interesting as holograms: they record better if you spray them with white or aluminium paint. However, there are much more exciting subjects. After a little practice with holographic composition you can put together scenes with ivory figurines, or glass animals, or paperclip sculptures. Make some 'abstract' compositions with parts of an old watch. Some of the most exciting effects

Pete's Dragon, a large real-image reflection hologram commissioned by Disney studios and made by Nick Phillips at Loughborough University of Technology. *Photograph by courtesy of Nottingham Evening Post.*

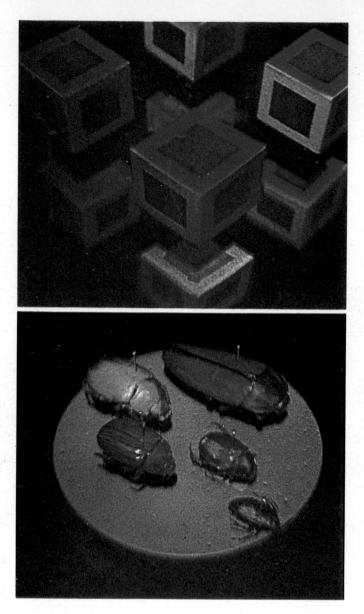

Two examples of one-step white-light transmission holograms. *Photographs by courtesy of P. Hariharan, National Measurement Laboratory, CSIRO, Sydney, Australia.*

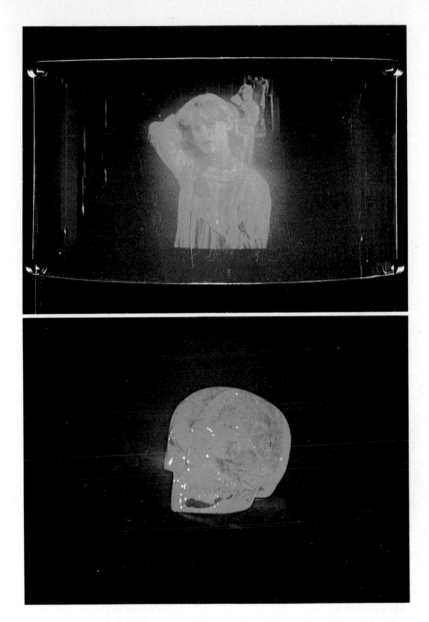

Above An example of a commercial integral hologram. These are sold complete with base containing the reconstruction light source. This hologram is by Nigel Abraham. *Photograph by Nigel Abraham. Below* Aztec crystal skull, laser transmission hologram by Nick Phillips. *Photograph by Theo Bergström.*

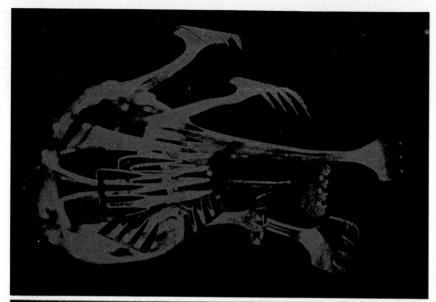

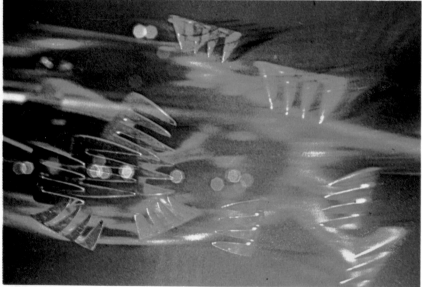

Two versions of the same subject, by Harriet Casdin-Silver.
Above 'Equivocal Forks I' (1977). Laser transmission hologram. *Photograph by Nisha Bichajion.*
Below 'Equivocal Forks II' (1978). White-light transmission hologram. *Photograph by Palumbo.*

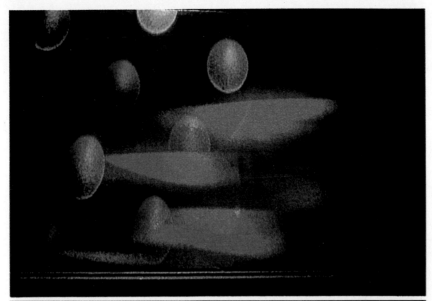

Detail from 'Twelve Milliwatt Boogie' (1979) by Rudie Berkhout, seen from two different viewpoints. The original is three white-light transmission holograms mounted as a triptych. It was made by multiple-exposure techniques, and combines virtual, image-plane and real images. The detail is from the left-hand panel. *Photograph by Rudie Berkhout*.

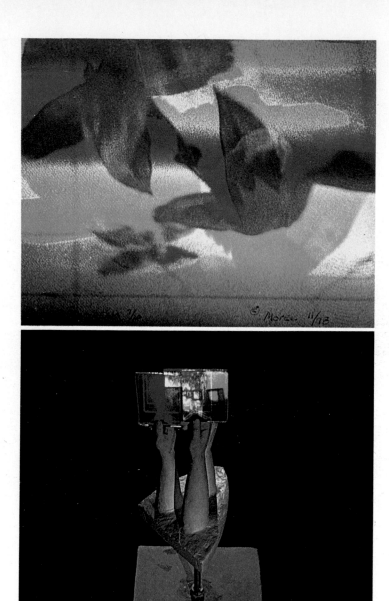

Two works by Sam Moree. *Above* 'Sidewalk Dreams' (1978), a white-light transmission hologram containing virtual, image-plane and real images, and combining shadow techniques with realistic images. *Photograph by Linda Law.* *Below* 'Triangle Hands' (1977), a multi-media sculpture including one laser and two white-light transmission holograms. *Photograph by Nancy Safford.*

Above 'Homage to Louis Comfort Tiffany' (1978) by Ruben Nuñez. A white-light transmission hologram which exploits the rainbow effect. *Photograph by Nigel Abraham. Below* 'Solar Markers' by Margaret Benyon. These reflection holograms are of underwater objects mounted on burnt rock and illuminated by the sun. This composition is intended to stress the elemental nature of matter.

Above 'Mandala' by Bill Reber. A 4 × 5 in white-light transmission hologram. *Below* Detail of 'Mandala', showing how the hues alter with changing viewpoint. *Photographs by Linda Law.*

are optical ones. Include a pocket magnifier, or a small phial filled with water, in your foreground. A diamond ring (or costume jewellery, if you do not run to diamonds) can produce a most impressive sparkle if you arrange it carefully. Do not attempt living flowers, however. In even the few seconds of exposure they will move enough to smear out the fringes and spoil the hologram. Dried or plastic flowers should be successful.

Viewing by partially coherent light

If you want to show your holograms to your friends (and who wouldn't?) there is no need to take your laser along. If there is a yellow sodium street lamp nearby (not the pinkish high-pressure type) it will do just as well. You can even view a hologram by the light of a desklamp if you cover it with a cardboard disc with a 6 mm ($\frac{1}{4}$ in) hole covered with a red photographic filter or a piece of stage gel. Although not as well resolved as with a laser, the image will still be fairly impressive.

Making a cosine grating

Using this setup you can make a cosine grating with a spatial frequency of about 1400 cycles per millimetre. This is twice as fine as a commercial diffraction grating. You need to replace the object with a second mirror directed so that the two beams overlap at the film plane. The two beams should have approximately the same intensity. If you use a glass plate (a slide cover glass will do) instead of a mirror, with a black card behind it, for the 'object', the balance should be about right.

Process the film (bleach processing is preferable). Mount the grating in a slide mount, masking off any unexposed portions. Cut a 2 mm slit in a card 50 mm square, put this in a slide projector, focus up, and place your grating immediately in front of the lens. You should see a white bar in the middle, and two beautiful rainbow-like spectra, one each side of it. If you are a teacher or lecturer, and are involved in teaching basic optics or colour theory, a good cosine grating can be an invaluable visual aid. By using filters of the primary and secondary hues in the beam you can provide a most elegant demonstration of the principles of additive and subtractive colour synthesis.

White-light reflection holograms

A few years ago it was announced that the Soviet Union was producing holographic records of many of its art treasures; a number of these have since appeared in exhibitions in other countries. Wherever they have been seen their incredibly lifelike quality has attracted admiration. They were made by one of the simplest methods of producing holograms, Denisyuk's method (see page 44). Because there is only one beam, which does duty for both reference and object beams, this method is usually known as single-beam reflection holography; apart from its basic simplicity, it has the advantage that the reconstruction beam need not be monochromatic.

Equipment required

The only requirements are the laser, a beamspreader and a float glass block (6 mm is best). It is important to use the highest-resolution film (e.g. Agfa 8E75HD), as the faster emulsions do not give sufficiently fine grain to record the very closely spaced fringes. This is normally supplied without antihalation backing. If your film is backed you will have to remove the backing before you use the film. You can do this by wiping the back of the film with a pad of cotton wool moistened with alcohol (ordinary methylated spirit will do, provided you carefully polish off any oily residue before using the film). This can be a tricky operation in total darkness; but you can use a panchromatic safelight with a bulb of double the normal wattage (but do not allow this to overheat). This light is also useful when cutting up 4 × 5 in or 8 × 10 in sheet film into smaller sizes, and for processing films.

Method of setting up

It is necessary with reflection holograms to have the reconstruction beam coming from about 45° above. As it is awkward to raise and tilt the laser and beamspreader, fix the objects to their plinth with 10-sec glue, and place the whole setup on its side. You can

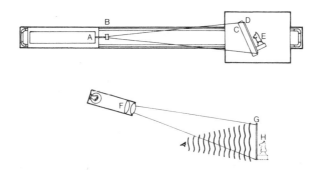

Making and viewing a single-beam reflection hologram. **A** Laser (shown mounted on optical bench). **B** Beam spreader. **C** Float glass plate angled at 30°–45°. **D** Film, emulsion outwards. **E** Objects (turned on side). **F** White reconstruction light. **G** Hologram. **H** Virtual image.

support the film between two pieces of glass, or simply tape it to a glass plate, with the emulsion facing the object.

Making a test exposure

You will need three pieces of film to make your initial tests. Make two steps of exposure on each, holding a piece of card to cover half of the film after the first part of the exposure. In this way you can make exposures of, say, 10 and 20 sec on the first film, 40 and 80 sec on the second, and 2½ and 5 min on the third (or about one-tenth of these times if you are using a 5 mW laser). Keep the card well away from the objects. Do not forget to mark the exposures on the back with a wax pencil before you process the films.

If you develop and fix a reflection hologram in the usual way you may be somewhat disappointed with the results. The silver layers absorb so much light, and so much is reflected directly back along the beam, that unless your reconstruction beam is very powerful, any image you may be able to see will be decidedly murky. Development followed by bleaching to produce a phase hologram is almost essential. The technique demands a very generous exposure, enough to produce a density of 3 or more. Using a 0.5 mW laser on a 2 × 2.5 in cut film, with the illuminated disc just covering the film area, the exposure will probably be upwards of a minute; about ten times that for an amplitude hologram. It is impossible to be more specific, as a beam that is nominally 0.5 mW power may in fact be anything between 0.1 and 1.0 mW. More powerful lasers are accurate in their specifications.

Processing

Process the film by one of the methods suggested in Appendix 3. You can develop the films in a dish, or load them into a 120 size spiral spool tank for processing, in which case you can carry out the whole operation in daylight. If you do use a spiral spool ensure the pieces of film are well separated, so that the agitation does not move them enough to cause them to overlap. When you come to 4 × 5 in films, you may find it easier to process them in film hangers in tanks.

Assessing the correct exposure

Examine the hologram against a dark background by the light of a slide projector, spotlight, microscope lamp or any small, powerful white light source above and behind you. You should be able to see the virtual image just behind the film. Depending on the method of processing you used, the image may be green rather than red. One of the stepped exposures should give a brighter, clearer image than either of the other two.

If there is little difference in quality between two adjacent exposures, say 40 and 80 sec your exposure should be 60 sec. Once you have got your exposure right, you should not need to alter it for the remainder of your holograms, unless you have objects which are significantly lighter or darker than your original test objects.

Viewing the real image

As with transmission holograms, there is a real image associated with the hologram, and it is on the other side of the film. You can see it if you turn the hologram round so that the emulsion side is towards you and change the angle of the reconstruction beam from above to below. Although, strictly, you should use a converging beam as you did for the transmission hologram, you will probably manage to get quite a good image with a slightly diverging beam, provided the original objects were close to the film and the reconstruction source is not too close. The image will appear to be standing out in front of the hologram. It will again be pseudoscopic and show reversed parallax.

An 'orthoscopic' real image

We saw on page 43 that one way of turning a pseudoscopic real image into an orthoscopic one was to use the pseudoscopic image as the object for a second hologram. For the beginner this is not a particularly easy task. However, there is an ingenious way round the problem. You simply start with a pseudoscopic object! You can make such an object very simply, by taking a negative cast or mould of an object, using flexible moulding material. There is a self-curing rubber material available from large education stationers and some model shops under the name of Supercast; you can also obtain flexible impression-making material from dental suppliers. Of the two, the modelling material is easier to use. The method is simply to paint layers of the material on the object until the mould is thick enough, then strip it off. As the mould is flexible it does not matter if the object has re-entrant surfaces. Cut it vertically close to its widest part, using a sharp knife or razor blade, so that you are left with the front half (or slightly less than half) of the object. Now spray the inside of the mould with white or aluminium paint. When this is dry, paint the cut edge and the outside of the mould adjoining the face with matt black paint. Set up the object and film as you would for a virtual-image

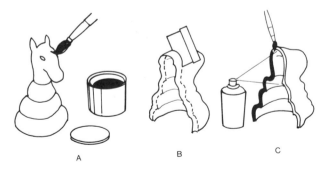

Making a pseudo-object. **A** Paint front half of object with several coats of flexible moulding compound. **B** Cut squarely down centre line. **C** Spray interior with aluminium aerosol, and paint cut-edge black.

single-beam reflection hologram, with as small a space as possible between the film and object. It is a good idea to fix the object directly to the film with a spot of rubber cement, provided you make sure all traces of it are removed before you start processing the film.

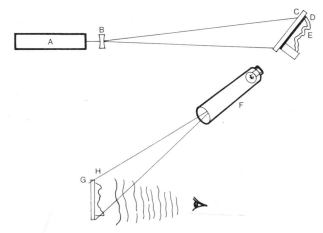

An orthoscopic image of a pseudo-object. **A** Laser. **B** Beam spreader. **C** Float glass block. **D** Film with emulsion outwards. **E** Pseudo-object. **F** White reconstruction light. **G** Hologram, emulsion side towards viewer. **H** 'Orthoscopic' real image.

There is, however, one important difference in the lighting. This time you must illuminate the object from below, not above, so offset the object and film so that the light appears to be coming from below. This change of lighting direction is necessary because you want the reconstruction light to come from above, and the hologram will be reversed.

As an interesting aside, on examining the object, it exhibits itself all the peculiar characteristics of the pseudoscopic real image, especially if you keep one eye closed. Put the object on a table, and stare at it, keeping one eye covered. It will seem to be out of its 'shell', in its normal convex form. Move your head slowly from side to side. You will get the same weird rotating effect as you do from a pseudoscopic real image. Try moving your head up and down and diagonally. The effect is particularly striking with human faces, which appear to take on a life of their own.

A wrap-round hologram

This is one method for making 360° holograms. It uses a piece of holographic film bent into a cone. In principle, this cone is placed over the object, and the single beam comes from directly overhead. In practice this awkward lighting arrangement is unnecessary: you simply fix the object to a wooden block using 10-sec glue or Blu-tack. You then fix the cone of film over it, turn the block on its side, and illuminate it horizontally.

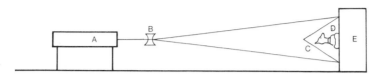

360° single-beam hologram. **A** Laser raised on block. **B** Beam spreader. **C** Film bent into conical shape or taped together into pyramid. **D** Object. **E** Block to which object and film are fixed.

To make the cone you need a semicircular template of 5 in diameter. Use it to cut a semicircle from a piece of 4 × 5 in film. (You can use the leftover piece for test exposures for other work.) Roll the film into a cone, emulsion innermost. You will have to overtighten it a little, then open it out. Butt join the edges with a narrow strip of cellulose tape.

As the diameter of the cone is only 40 mm (1.6 in) and the angle of the cone is 60°, you will not be able to have an object larger than about 25 mm (1 in) cube. However, a tiny ballerina, a pair of dice or a small glass animal are very effective. You can make a slightly larger cone from 70 mm film.

Your wooden base should be sprayed with aluminium or white paint. Lay a ring of Blu-tack round it for fixing the film in place, and flatten the ring down. Put the cone firmly in place; put the block on its side, and give the setup 15 min to settle down and release the stresses in the film. When you have made the exposure, open out the cone for processing.

Rebuild the cone when the film is dry, and fix it with contact adhesive to a black plastic or cardboard base. The illuminating light can be an ordinary 100 W bulb, as long as it is directly overhead; and you should be able to see the virtual image from any angle.

You can also make conical holograms for mounting on vertical surfaces such as walls. For these you need to mount the object so that it is erect when you turn the block on its side.

When you first set up the cone of film, try to arrange the join to be in an unimportant place, for example at the back of the model, or underneath, if the hologram is the horizontal variety.

71

You can make tetrahedral and pyramidal holograms by this method, too; and as the film is not bent you can expose these without having to wait so long for the film to settle down. You can make these rather larger: a 4 × 5 in film gives a slant height of 5 in and a base length of nearly 4½ in.

Cutting film to make a tetrahedron or pyramid. A pyramid has a greater base area than a tetrahedron, but is slightly lower.

If you have managed to produce good holograms using these small-size arrangements, you may find your appetite whetted for bigger things. Larger holograms give better opportunities to spread your creative wings, and they are inherently no more difficult to make. You need a different way of supporting your optical components in this larger format; the next chapter shows the method. This also permits the making of some exciting types of hologram we have not met so far.

Larger holograms with a sandbox

Why a sandbox?

One of the difficulties about pictorial holography is that the image is the same size as the object. This means that if we want larger compositions we need larger films, and this in turn means that the scale of our setup has to be larger. Some years ago Gerald Pethick pioneered a technique which, because of its versatility, cheapness and above all, reliability, has become one of the most popular for larger-scale holography. It employs a large rectangular box filled with sand. The optical components are fixed to posts, which are simply pushed into it. The sand makes a good damping medium, and its heavy weight provides the inertia required for isolation from vibration when the box is supported on inner tubes. As it can be removed, there are no problems about resiting the box should this prove necessary.

Choosing a site

The best site is a room at ground level or in a basement, with a concrete floor. Wooden floors are subject to vibrations, and may not stand the considerable weight of the sandbox. If you don't own a house with a basement, a garage is the next best thing. You will need to black out the windows, of course. It is also important to exclude draughts, and to avoid having sources of convection currents such as radiators and refrigerators in the same room. Water pumps are a source of noise and vibration, too. Turn off any radiators some time before you start work; and if your sandbox has to share houseroom with your freezer, switch that off, too. As long as you leave the lid shut, the contents will be all right for at least four hours.

Constructing a sandbox

The box should be as large as you have space for. The absolute minimum is 1 m (39 in) square; but the size and range of the work you can do with such a small box is very limited. The optimum size is approximately 1.5 × 2 m (5 × 7 ft), and that is the size described here. The most suitable material for the box is tongued and grooved floorboarding, thoroughly dried out. The width of the boards (excluding the tongue) is 100 mm (4 in); so when you have decided what is the largest size box you can fit into your

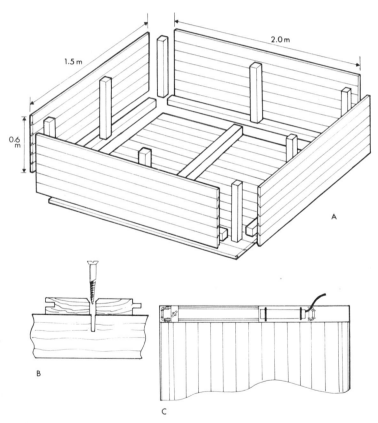

Constructing a sandbox. **A** Method of construction, using tongued and grooved floorboarding and 50 mm square batten. **B** Method of fixing boards to batten. Note direction of grain. **C** View from above showing optical bench with laser and relay mirror. Direction of boards is across the shorter dimension.

floor space, allowing free movement round at least three sides, work out the number of widths you will need for the floor of the box. The boards will be secured to 50 mm (2 in) square battens.

The sides of the box should be six boards deep. To make a neat job the end boards should overlap the side boards; so cut the side boards to the same length as the floor boards, and cut the end boards 36 mm ($1\frac{1}{2}$ in) longer than the width of the floor. Leave this last job until you have completed the floor. Take the boards that are going to form the floor, and lay them across three battens. When you examine a board you will notice that the tongue is not central, but is displaced slightly to one side. Position the boards with the smaller rebate uppermost. The annual rings of the timber will then have their convex side downwards. These rings tend to straighten as the wood ages, resulting in a warped surface. If this warp is concave to the battens, and you have fixed the boards by means of screws through the mid-line, it will be minimized. When you are sure you have all the boards the right way up, knock them together using a wooden mallet or rubber-faced hammer.

When you come to fix the boards to the battens you will find it difficult to hold them tightly enough together. The best way to do this is to use a pair of sash cramps. These operate like a carpenter's vice, but with the jaws much farther apart. If you do not have any friends who are handymen, you can hire large sash cramps quite cheaply from large do-it-yourself centres. If the available cramps are too short, you can bolt two together to get one of double the length. If you get a friend to help you it is quite easy to cramp the boards at both ends simultaneously. You won't be able to avoid bruising the outermost tongue when you tighten the cramps, but this does not matter, as it will eventually be removed.

Your next job is to drill and countersink the holes ready for the screws. Using a 5 mm ($\frac{3}{16}$ in) drill, make holes right through the boards on their centre lines where the battens are to go. Now line up the battens and, using these holes as a guide, drill a short distance into the batten, using a 2.5 mm ($\frac{3}{32}$ in) drill. Countersink the holes (if you do not possess a countersink bit, use a 12 mm ($\frac{1}{2}$ in) drill bit to make a shallow depression).

Use 44 mm ($1\frac{3}{4}$ in) No. 10 countersunk woodscrews. As you are going to put in a total of 136 screws before the box is finished, you would be well advised to use a method that takes the minimum of effort. Scrape the screw across a damp bar of soap before inserting it in the hole, and hammer it in like a nail until only about 6 mm ($\frac{1}{4}$ in) projects. Then tighten it up with a ratchet screwdriver. It sounds a bit cavalier; but in fact it will not do the slightest harm to the timber if you have predrilled the holes correctly.

The outermost screws for the centre batten only should be inset about 25 mm (1 in), as this batten has to be cut short to allow for the edge battens that the sides are going to be secured to. When you have fixed all the boards to all three cross-battens, remove the sash cramps. Fit the two longitudinal battens, which you should cut 100 mm (4 in) short to allow them to fit between the outermost cross-battens. Secure these in the same way as the cross-battens, with seven screws set 275 mm (11 in) apart, measuring from the centre outwards. Turn the floor over.

Build the two long sides in exactly the same way. This time, however, all three cross-battens should be 50 mm (2 in) short at the 'groove' edge and flush at the upper, 'tongue' edge. You will, of course, not need a longitudinal batten, as you have already fastened this to the floor.

Position one of the sides against the floor, and get a friend to hold it upright while you drill holes near each end and at the centre. Offset the holes sufficiently to miss the screws that are already in the batten. Insert three screws. These will hold the side in place while you fix the other side similarly. Next, cramp one set of end boards together without using any battens. Raise one end of the floor so that the cramp jaws will be clear of the ground, and align the end boards. Drill and fix the two end screws to the bottom batten, then align the sides and fix the top board at each end. Fix the third or fourth board down, similarly. Do this for both ends. Remove the cramps and fix the rest of the end boards. Now put in the remaining four screws into the longitudinal battens each side, and the remaining two screws into the end battens. Remove the tongue from the topmost boards with a chisel or jack plane, and smooth the top with glasspaper. Your box is now complete, though you may like to add a metal tie-rod across the shorter dimension near the top, to give added bracing.

Supporting the sandbox

The next item is a support for the box. Heavy-duty Dexion or other angled perforated strip is a good material to use. You can get this from engineering suppliers and large do-it-yourself centres. The dimensions of the frame are 90 × 150 cm (approx. 3 × 5 ft) × 30 cm (approx. 1 ft) high, and there should be an extra brace in the middle. You therefore require six pieces 30 cm long, four 15 cm, six 90 cm, 24 gusset plates and 80 bolts and nuts; that is, one gusset plate and three bolts and nuts for each corner, and eight bolts and nuts for the centre stays.

If possible, the support should be partially isolated from the floor. The best method is to stand it on up to six layers of foam-backed carpeting. If you have not laid any carpets recently, you can get samples of discontinued lines of carpeting from carpet dealers or market stalls for a few pence each.

The isolating support for the box itself is six scooter tubes size $3\frac{1}{2} \times 10$ in (or a similar size). These must be new tubes. They will be seated on a piece of chipboard or blockboard about 1.2×1.8 m (4×6 ft) in size. In order to be able to get at the valves for re-inflation when necessary, you will need to cut circular holes about 15 cm (6 in) in diameter, one for each tube.

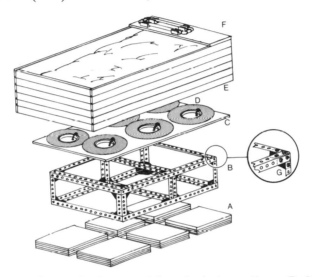

Supporting a sandbox. **A** Squares of foam-backed carpeting. **B** Heavy duty Dexion frame. **C** Blockboard support with holes for access to valves. **D** Scooter inner tubes. **E** Sandbox. **F** Support for laser and relay mirror. Inset: corner of frame showing gussets, **G**.

Place the board on the support frame. Inflate the tubes until they are resilient but not hard (about 6 psi) and place them on the board with the valves in such a position that you can reach them from underneath. If you are short of space or help you may find it easier to jack up the box on bricks and position the carpeting, frame, board and tubes under it before removing the bricks. Line the bottom of the box with polythene sheet, which you can get in 2 m width from garden centres.

The sand

Your next task is the sand. For a 1.5×2 m box you will need 1.8 cubic metres (50 cubic feet), or 2 tonnes if sold by weight. You can get suitable sand from a builders' merchant. It is known variously as dry washed silica sand, washed river sand, or yellow building sand. It will usually be dumped in your driveway; and you will have to move it in buckets. Do not hurry over this: apart from avoiding the possibility of injuring yourself, you need the sand to be dry. Skim off the surface as the sand dries, and spread it thinly over the box area. Take several days over this, and cover the heap with a polythene sheet if it starts to rain. If you live near the sea, you can use fine sea sand and save yourself a few pounds; but you will need to wash the sand and spread it out to dry thoroughly first, as otherwise the mineral salts remaining in it will absorb moisture so that the sand becomes sticky and the box walls and floor damp.

Supporting the optical components

The best support for the laser and relay mirror is the optical bench used described on pages 58–60. If the width of your sandbox is greater than the length of your existing baseboard, replace this by a piece of floorboarding of length equal to the width of the box. Fix the rails to this board towards one end, so that the relay mirror can be at the corner of the box to give maximum throw to the beam. You will already have made a support for the laser (page 59), but you will need to make one for the relay mirror which allows it to be moved along the rails or rotated, but is nevertheless firm and rigid. Make the support pillar from 16 mm (in) dowel or nylon rod, and the base

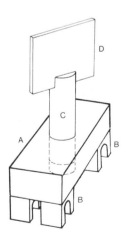

Support for relay mirror. **A** Wood block. **B** Pipe clips. **C** 16 mm dowel or nylon rod (fairly tight fit in hole). **D** Front-surface mirror.

from a 100 mm (4 in) length of batten, bored to take the pillar as a tight push-fit. Make sure you bore the hole absolutely square to the batten. Rub the inside surface with a piece of candle before you insert the pillar, to prevent it from seizing.

The remaining components will be on supports pushed into the sand. The best material for the supports is 38 mm (1.5 in) diameter plastics piping, which you can buy for home plumbing jobs from do-it-yourself suppliers. The piping needs to be of such a length that it almost reaches the bottom of the sandbox when the optical component is at the same height as the relay mirror. For boxes six boards deep, the piping should be about 60 cm (24 in) long. The component holders, which may be bulldog clips, Terry clips or lumps of Blu-tack on a table, should be fixed to 20 cm (8 in) lengths of 15 × 36 mm ($\frac{5}{8}$ × 1$\frac{1}{2}$ in) batten with the edges slightly rounded off. You can wedge these into the pipes.

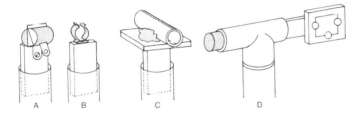

Mountings for optical components. **A** and **B** Mountings for mirrors, beam splitters and lenses. **B** Wooden batten is a tight fit in plastic piping. **C** General-purpose table with Blu-tack. **D** Overhead mirror with counterbalance.

This task is easier if you use a carpenter's vice to flatten the pipe slightly while you insert the batten. Leave about 50 mm (2 in) above the end of the pipe. You can, of course, remove the batten if you want to change to a different length of pipe. You will need longer pieces of piping for overhead optical components, and shorter lengths for low ones. For overhead mirrors with horizontal-axis movements use a two-way pipe connector, with the mirror fixed to a batten pushed into one arm and a counterweight in the other. Always counterbalance any asymmetrical components.

Interferometric steadiness check

Before you start making holograms you must be quite certain that your setup is free from any disturbances which could mar your results. To check this you need to set up a pair of mirrors and a beamsplitter to form the optical configuration known as a *Michelson interferometer*. The method is as follows:

1 Switch on the laser and adjust the relay mirror so that the beam is reflected across the box at an angle of approximately 45°.
2 Set a mirror close to the side of the box so that it reflects the beam back along very nearly (but not exactly) the same path, i.e. the two spots almost coincide at the relay mirror (if they do coincide exactly, the laser output may become unstable).
3 Insert a beamsplitter vertically in the light path, at approximately the centre of the box and at about 45° to the beam. A plain piece of float glass will do, though a 1:1 metallized beamsplitter is preferable.

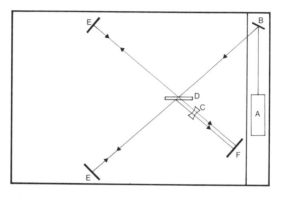

Arrangement for interferometer test. **A** Laser. **B** Relay mirror. **C** Beam spreader. **D** Beam splitter (50%). **E** Mirrors. **F** Screen. For clarity, the beam is shown as if undiverged.

4 Set a second mirror by the side of the box adjacent to the relay mirror, the same distance (to within a few millimetres) from the beamsplitter as the first mirror, so that the beam from the beamsplitter falls on it.
5 Set a screen at the remaining corner of the box opposite the second mirror, so that the beams from both mirrors appear as spots on it.
6 Adjust the first mirror and beamsplitter by successive approximations until the beam reflected from the second mirror via the beamsplitter forms a spot on the relay mirror which coincides with the spot produced by the first mirror, and the spot on the screen formed by the first mirror (via the beamsplitter) coincides with the spot produced by the second mirror (this sounds more difficult than it is). Mark the spot on the screen with a pencil.

7 Insert a beamspreader lens between the beamsplitter and the screen, and adjust it until the spread beam is centred on the mark on the screen.

You should now be able to see a set of black and red fringes. Ignore any stationary fringes, which are caused by double reflections in the beamsplitter: it is the moving ones that matter. They will probably move around for a few minutes until the sand has settled down, then become still. If your isolation is satisfactory, the fringes will remain stationary as long as everything remains undisturbed. If they shimmer in a regular kind of way, there is vibration present. If the movement is irregular, the cause is probably convection currents.

Once you have managed to eliminate any cause of unsteadiness of the fringes, see how much disturbance you have to create in order to make them move. Rap on the side of the box. Tap the support frame with a ruler. Stamp on the ground. Breathe into the space between the beamsplitter and one of the mirrors. Talk to one of the mirrors. Shout at it. Clap your hands at varying distances from it. Use a piece of card to make an 'exposure'. Wave the card about. In each case make a note of the intensity (if any) of the effect, and the length of time it takes for the fringes to settle down again. These tests should give you a good idea of what you can, and cannot, get away with in terms of disturbance. Remember, during the exposure, any movement of more than one-quarter of a fringe will seriously affect the quality of your holographic image; movement of a whole fringe will completely erase it.

You can also use this interferometer to find the coherence length of your laser beam. Starting with the optical path lengths equal, move one of the mirrors along the beam until you can no longer see the fringes clearly, then measure the difference in distance between the two mirrors and the beamsplitter. This is equal to half the coherence length, and the coherence length is the depth of field you have available. For a 5 mW laser it is usually of the order of 15 cm (6 in). You should repeat the interferometer test every time you are going to hold a holographic session. Always record the initial settling time: it is the minimum time you will have to allow between inserting the last optical component (usually the filmholder) and making the exposure.

The interferometer test is, of course, not confined to sandbox setups: it is well worth carrying out before you do any kind of open-table work.

Partial reflection at a beamsplitter

The proportion of light reflected at an unsilvered glass surface varies with the angle of incidence. For ordinary white light it remains fairly constant as the angle of incidence is increased from 0° to about 40°, after which it begins to rise more and more rapidly, reaching almost 100 per cent at grazing incidence. Linearly polarized light follows one of two curves which differ

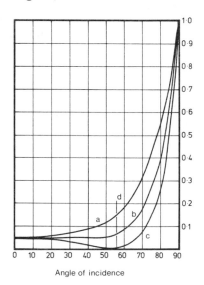

Reflectance vs angle of incidence for a glass surface.
(a) Polarized parallel to surface. (b) Randomly polarized. (c) Polarized perpendicular to surface. (d) Brewster angle.

considerably, depending on whether the polarization is parallel to or perpendicular to the plane of the reflecting surface.

Small lasers with random polarization, when used with plain glass beamsplitters, produce a reflected beam the intensity of which fluctuates slowly with time, as the orientation of the plane of polarization wanders. At an angle of incidence of about 56° its intensity may be anything from zero to about 15 per cent of the total intensity, and the reference beam : object beam intensity ratio can vary wildly. For this reason it is difficult to produce high-quality work with a randomly polarized beam. The insertion of a polarizing filter helps, but although the reflectance ratio will then be fixed, the total intensity will fluctuate and the exposure will be uncertain. It is much better in the long run to use a linearly polarized beam. If the plane of polarization is vertical (the usual alignment in lasers), the reflectance at 45° will be approximately

82

10 per cent; at 60° it will be about 20 per cent; and at 75° about 30 per cent.

Many arrangements call for a beamsplitter with a 1:1 reflectance ratio; and here it is impracticable to use plain glass. Appendix 2 gives addresses where you can obtain partially reflecting metallized glass for beamsplitters. These are not sensitive to differences in polarization, nor to changes in angle of incidence, though the coating absorbs about one-third of the incident light. You can also obtain variable-ratio beamsplitters from suppliers of optical components. These work on a different optical principle, and are very expensive when compared with plain glass or partially reflecting mirrors.

A beamspreader with a spatial filter

In larger holograms any non-uniformity of the reference beam is not merely an irritation: it can lead to patches of incorrect exposure and consequent degradation of the image quality. You can do a good deal to minimize this problem by ensuring that all optical surfaces, especially those of the beamspreader and the

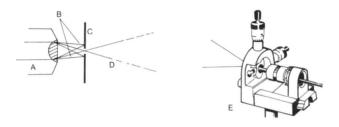

A pinhole as a spatial filter. **A** Microscope objective. **B** Diffracted beams blocked off. **C** Diaphragm with pinhole allows only zero-order beam to be transmitted. **D** Beam now carries no information and is completely 'clean'. **E** Commercial spatial filter.

laser window, are scrupulously clean. However, if you want an absolutely 'clean' beam you need to modify the beamspreader design. In general, when used as beamspreaders, concave lenses seem to give less trouble than convex lenses. However, the best type of beamspreader is undoubtedly a ×40 microscope objective (×60 for the smallest sandboxes) with a very small pinhole at its principal focal point. This arrangement gives a beam which is remarkably free from blemishes. The pinhole is called a *spatial filter*.

A brief explanation of this is that the focus of the objective is

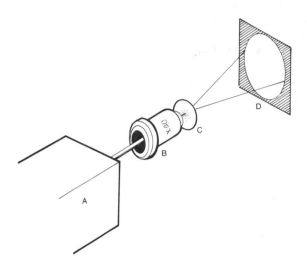

Eliminating 'noise' with a spatial-frequency filter. **A** Laser. **B** Microscope objective. **C** Pinhole at principal focus eliminates wavefronts diffracted by dust etc. **D** 'Clean' disc of light.

not quite a true point. It is a diffraction pattern with a central disc about 7 micrometres (μm) in diameter combined with random diffraction patterns caused by dust and dirt. If the diameter of the pinhole is sufficiently small to hold back all these diffraction patterns (which are outside the central disc) the result will be a completely clean beam. At the focal point of a microscope objective, about 80 per cent of the total light is contained in the central disc, and very little light intensity is lost by spatial filtering. With a simple convex lens the proportion is smaller; but you can still get a considerable improvement without losing too much light intensity.

Making a pinhole

Pinholes from suppliers of optical equipment are very expensive. The pinholes made for electron microscopes are much cheaper, and just as good. For a ×40 objective the correct diameter is 10–12 μm (sometimes micrometres are called microns), though a 25 μm pinhole is almost as good and much easier to position. Before deciding to buy a pinhole you may like to experiment with a home-made one. Take a small piece of aluminium foil such as a milk-bottle top or part of a frozen-pie container, place it on a sheet of glass and prick it with a small needle, the finest you can find. Press it down on the foil, twirling it lightly with your fingers. Do not push the needle right through. Only the extreme point should penetrate the foil. Examine the foil against a strong light to make sure there really is a hole (it should be almost invisible); then fix it to a metal washer about 12 mm in diameter with contact adhesive or cellulose cement, and trim the edge with scissors.

Alignment of pinhole

The pinhole needs to be very accurately aligned in all three dimensions, and this can be a tedious and frustrating business if your 'mount' is simply a ring of Blu-tack holding the washer to the front of the objective. Nevertheless, it can be done; and this method has the advantage of costing nothing. In the end, though, you will probably decide to get an accurate pinhole and an adjustable holder. Companies who supply lasers also sell spatial filters as complete items, but these usually cost almost as much as the laser. Ordinary optical mounts with two-way micrometer adjustments are a good deal cheaper, but still expensive. Suppliers of optical bench equipment sell component holders or 'saddles' with two-way vernier adjustments, and these are more reasonably priced, though somewhat less easy to use. If you have access to metal-working facilities you may be able to make your own adjustable holder, or modify an old vernier microscope stage. The method you eventually decide on thus depends on either your pocket or your skill and patience. Whatever method you choose, the procedure for alignment is the same.

Set up your microscope objective on its stand so that the laser beam passes squarely through its centre. Mount the pinhole about 1 mm from the surface of the objective lens, centred as accurately as you can by eye. Set up a vertical screen about 1 m (39 in) away, and darken the room. There will probably be a faint patch of light on the screen. If not, move the pinhole horizontally and vertically by turns until a patch appears. As you move the pinhole you will notice that the patch becomes fainter, and that it moves in the same direction as the pinhole. Now carefully move the pinhole nearer to the objective, watching the patch; when it becomes faint, stop. Adjust the pinhole again horizontally and vertically until the patch is at its maximum intensity. It should be somewhat brighter than before. Again, move the pinhole closer to the objective and readjust its position for maximum brightness. Eventually you will reach a point where the patch is very bright, completely free from swirls and rings, and does not move as you move the pinhole, but simply disappears and reappears. You have now found the precise position of the focal point, and your task is complete.

Unless you have micrometer adjustments, or high-quality verniers free from backlash, you will probably not be able to work with a pinhole of diameter less than 25 μm; but such a pinhole will usually give almost as good a beam as the theoretical minimum. In fact, even larger pinholes of up to 100 μm diameter will still bring about a considerable improvement.

Once you have aligned the pinhole correctly it should not need further adjustment. The only thing you have to take care over when setting up is that the axis of the objective is in exact alignment with the laser beam and that the beam strikes the objective lens centrally (otherwise the patch of light will be displaced).

Object-illuminating beamspreaders

For these it is not so important to have an absolutely clean beam, nor need the divergence angle be small. To spread the beam without changing its direction use two concave lenses: the first should have a focal length of about 6 mm; the second of about 30 mm. Increasing the distance between the lenses will increase the angle of spread. You can mount both lenses on the same post, using Terry clips or Blu-tack.

Alternatively, you can combine the functions of beamspreader and mirror. For this an ordinary ballbearing is ideal. The surface finish of a new ballbearing is at least as good as that of any high-quality optical component (the maximum allowable roughness is less than 50 nm); and it is cheap, tough and easy to keep clean. The focal length of a ballbearing is one-quarter of its diameter, so that if you want say, 6 mm focal length, you will require a diameter of 24 mm. The diameter of ballbearings is sometimes specified in fractions of an inch rather than millimetres, so it may be advisable when buying them to bear in mind that 1 in is the same as 25.4 mm; 24 or 25 mm is about the largest useful size; the smallest is about 6 mm. You can get ballbearings from suppliers of engineering equipment or from most garages. Keep them, well separated, in a box lined with cotton wool, and never touch them with your fingers. Pick them up with a paper handkerchief or a piece of lens-cleaning tissue. You may sometimes find that a ballbearing produces an intricate pattern instead of an even illumination: if so, use it in conjunction with the diffusing reflector described below. To produce a more diffused, soft light, interpose a matt glass between the beamspreader and the object. The farther from the spreader it is, the greater will be the degree of diffusion. If you use two matt glasses a centimetre or so apart, you can dispense with the lens and the illumination will be more diffuse still. You can mount these on a single post with a bulldog clip each side. With some lighting arrangements it is better to use a diffusing reflector rather than a matt glass. This also has the advantage that you can modify the intensity of the illumination across the object by curving the reflector. To make such a reflector, cut a piece up to 15 cm (6 in)

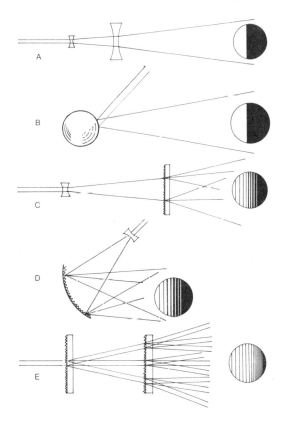

Object illuminating beam spreaders. **A** Two concave lenses. Spread depends on focal length and separation of lenses. No diffusion. **B** Ballbearing. Spread depends on diameter. No diffusion. **C** Concave lens and matt glass. Moderately diffused. **D** Aluminium sprayed reflector can be curved to control intensity and direction of illumination. **E** Two matt glasses give maximum diffusion.

square from a used food tin, spray it with aluminium paint, and mount it on plastic piping using one of the methods described earlier.

Transmission holograms

The setup for these is simple and straightforward, and a good one for learning basic sandbox techniques. The arrangement is basically the same one as you used for your small transmission holograms, but you can use larger objects, and 4 × 5 in film. In addition to the relay mirror you will require a plain glass beamsplitter, a mirror and a ballbearing about 12 mm (0.5 in) in diameter. The best type of filmholder is a cut film slide as used in small studio cameras. Mount it using a large bulldog clip fixed to a plastic pipe. You can then remove it for loading and unloading with the minimum disturbance to the setup. You also need a sheet of heavyweight black card to block any stray light from the object-illuminating beam from reaching the film. You can fix this to a stake; or you can cut it large enough to be pushed directly into the sand.

You may have noticed that your small transmission holograms contained a pattern of dark and light vertical bars about 0.5 mm wide. This effect was caused by interference between the primary (front surface) and secondary (rear surface) reflections from the beamsplitter; while this does not have a very serious effect on the quality of the image, it is better to be without it. If you use 3 mm (or thicker) float glass for the beamsplitter, the primary and secondary beams will be separated sufficiently for you to be able

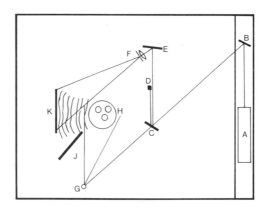

Sandbox layout for transmission hologram with one illuminating beam. **A** Laser. **B** Relay mirror. **C** Plain glass (5–10% reflectance) beam splitter. **D** Secondary reflection block. **E** Mirror. **F** Beam spreader with spatial filter. **G** Ballbearing. **H** Object. **J** Black card light baffle. **K** Film. *Note:* in this and subsequent diagrams, the size of the object and film, and the angle of beam spread, have been exaggerated in the interests of clarity. Other measurements are approximately to scale.

to block off the latter by inserting a wooden stake, painted matt black, in its path.

Set up the relay mirror first, adjusting it so that the beam goes diagonally across the box. It should cross the side wall about three-quarters of the way down towards the end. Position your object on the mid-line of the box, roughly opposite this point. Position the ballbearing so that the object is uniformly illuminated. Now arrange the beamsplitter and reference beam mirror so that the paths of the reference and object-illuminating beams form roughly a parallelogram. Insert the reference beamspreader (with a spatial filter if you have one). Position the filmholder roughly parallel to the object-illuminating beam, and arrange the reference beam so that it illuminates the film area uniformly. Now use a piece of string to check that the optical paths of the two beams are equal to within about a centimetre. Measure the object-illuminating beam to the centre of the front part (the visible part) of the object; and do not forget to add the distance from the object to the film. Next, insert the black card light-baffle between the film and the ballbearing so that no direct light from the object-illuminating beam can reach the film. Now, with the black stake, block the secondary reflected beam from the beamsplitter. Measure the relative intensities of the object and reference beams, using an exposure meter, with each blocked off in turn. Measure the object beam with the meter pointed at the object, and the reference beam with the meter pointed at the reference beam spreader, straight down the beam. The intensity ratio at the film should be 3:1 or slightly higher; but as the reference beam is striking the film obliquely its effective intensity is lower than the measured intensity, and will be correct when the meter readings give a ratio of 4:1 or 5:1.

Make any necessary small adjustments to the beam intensity ratio by moving the reference beamspreader towards or away from the film, or by changing the size of the ballbearing. In this way you can usually avoid having to place neutral-density filters in the reference beam unless your object is particularly dark. Make a note of the meter readings for the separate beams and keep this for future reference.

Some makes of exposure meter are not sufficiently sensitive to give a reading at the low intensities characteristic of these larger setups. If your meter does not give a usable reading, you would be well advised to get a photometer rather than to buy a more expensive meter; it can be as simple as a photodiode connected to the 'ohms' terminal of a multimeter. A more sensitive meter is based on a phototransistor and amplifier. Circuit diagrams, components and construction kits are advertised in home electronics magazines. You can also obtain photometers ready-made from suppliers of electronic meters.

Now load your film in preparation for making the exposure. Cut off the laser light with a black card, put the film in position, and allow the sand to settle down before you make a test exposure. After you have inserted the last component in the sand it will continue to settle for several minutes. You will have an estimate of this time from your interferometry check.

If you made a note of your meter readings when you were making your small transmission holograms, you will be able to estimate the exposure reasonably accurately. If not, you will have to make a series of exposures as described on page 55. The correct density for an amplitude (unbleached) hologram is about 0.6–0.8 (a transmittance of between 25 and 16 per cent) and for a phase (bleached) hologram about 1.5–2.0 (3–1 per cent). You can check these values using an exposure meter or a photometer, as described on page 56. Measure the phase hologram before bleaching. You can avoid fogging the unfixed film if you immerse it in an acid stopbath and give it a thorough rinse before measuring its density.

Your lighting effects will have much more freedom if you use two or more object-illuminating beams, and diffusers. The object-illuminating beamsplitter should be of the metallized type with a beam ratio of 1:1, and should be tilted so that it deflects one beam upwards; by using an elevated diffusing reflector, you can direct one of your beams at the object from above, giving a more natural modelling effect. In fact, you can use any of the

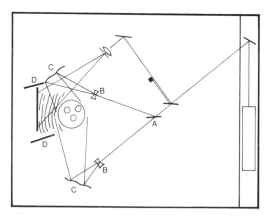

Layout for transmission hologram with two illuminating beams, using diffusing reflectors. Components additional to, or differing from, those in previous diagrams are as follows: **A** Metallized (1:1) beam splitter. **B** Concave lens beam spreaders. **C** Diffusing reflectors **D** Black card light baffles.

90

lighting arrangements suitable for table-top photography, with diffused beams playing the part of floodlights, and undiffused beams that of spotlights. Remember, though, that in holography you need to compose your picture in three dimensions, not two; your lighting needs to be attractive when seen from all the possible points of view within the hologram window. And you must remember to check that all the optical paths are equal.

Reflection holograms

In a reflection hologram the reference and object beams strike the film from opposite sides, the emulsion being towards the object. For this type of hologram you therefore need to support the film between two glasses. Tape these together down one side to form a hinge. You can use the same bulldog-clip holder as for the opaque filmholder. The reference beam needs to be offset

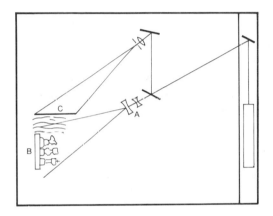

Reflection hologram with single illuminating beam. **A** Pair of concave lenses. **B** Object mounted on its side. **C** Reference beam strikes back of film at 45° incidence.

upwards at an angle of incidence of 45° or more. With only one object-illuminating beam, the simplest way to do this is to operate the whole optical layout, including the object, on its side. This avoids the need to elevate the reference beam.

The sequence for setting up is the same as for transmission holograms, though, of course, the route taken by the beams is different. The intensity ratio of reference to object beam should be lower, between 1:1 and 2:1.

You should always bleach-process reflection holograms, as this gives a much brighter image. A number of formulae and methods

91

are given in Appendix 3. The best results seem to be obtained by exposing to give a very high density, 3.0–4.0, corresponding to a transmittance of 0.1–0.01 per cent. Do not make up the bleach until you are fairly satisfied with your setup, as in general bleach baths deteriorate quickly after mixing. To carry out your initial tests simply develop the film without fixing, immerse it in an acid stopbath, wash it for 5 min and dry. You can examine the developed film by the light of the sun or a projector lamp. The image will probably not be very strong; but it should be visible. Once you are satisfied that a holographic image is present, go ahead and make up your bleach bath. Try it first on the film you have already developed. You will probably find that the best image is given by a hologram that had a good deal more exposure than the one you chose when they were unbleached.

Depending on your processing method, the image may be orange, yellow or green. The reason for the change in hue is that the emulsion has shrunk slightly during processing, and the separation of the layers is now equal to the half-wavelength of yellow or green light instead of red. There are various swelling agents to restore the correct hue: details are given in Appendix 3.

Reflection holograms are best mounted between two glasses and backed with black velvet or flock paper. Many people prefer to use holographic plates, which, of course, do not need mounting either for exposing or viewing. However, plates are more expensive than film, and are not easy to cut up for test exposures.

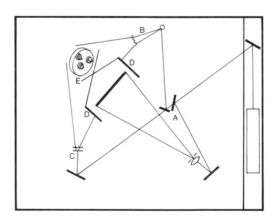

Reflection hologram with two illuminating beams. **A** 1:1 beam splitter. **B** Main illumination with ballbearing and single matt glass. **C** Fill-in illumination with double matt glass. **D** Baffles. **E** Object and film tilted forward about 20°–25°. The reference beam is also elevated 20°.

If you do use plates, you may have to remove any antihalation backing before exposing them. You can do this with cotton wool dipped in alcohol. Make sure you remove any oily residue before exposing.

With a single object-illuminating beam the lighting is somewhat restricted. As with transmission holograms, dividing the beam gives more opportunities for creative lighting effects. With this arrangement it is difficult to illuminate the object satisfactorily if it is on its side, and you will probably find it more practical to have it upright but tilt it forward at an angle of 20–25° from the vertical, and tilt the reference beam by about the same amount.

Focused-image holograms

We have not previously met this type of hologram. Although it is a true hologram, it requires a lens. A focused-image hologram produces an image that is in, or close to, the plane of the film. It can be virtual or real, or, if the image straddles the plane of the film, partly real and partly virtual. It is orthoscopic, and you can view it by the light of an ordinary lamp bulb. Also, the image is *achromatic* (not strongly coloured). The disadvantages are restricted depth of field and parallax.

The lens is used to produce a real image of the object in, or close to, the plane of the film. By adjusting the object–lens–film distances you can produce a magnified or diminished image. It looks as if the rule about a holographic image always being the same size as the object has been broken; but in fact it is the optical image that forms the object, and this is the same size in the hologram as it was originally. The amount of parallax available depends on the f number of the lens; so if you possess an $f1.5$ or $f2$ camera lens, this will do very well for objects not more than an inch or two across. For larger objects you will need a larger diameter lens; you can use a condenser lens from an enlarger, or make you own liquid-filled lens pages 106–7.

Focus the image initially on a piece of card. If you want a virtual image, place the film slightly farther away: if you want a real image, place it slightly nearer. Do not overdo this. If you have the 'object' more than a few millimetres from the film plane it will not reconstruct in white light. As the 'object' is inverted, you will have to re-invert the hologram for viewing, and when you do this you will need to switch the reconstruction beam to the other side as well. You can view focused-image holograms from the reverse side, and just as with conventional transmission holograms, you will get a pseudoscopic image. You will also get a

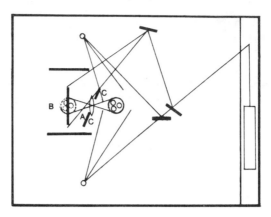

Focused-image transmission hologram. **A** Lens. **B** Focused image in plane of film. **C** Baffles to prevent direct light from object reaching film. *Note:* although this layout looks cramped, this is because of the exaggerated object and film size. In fact, there is plenty of space.

pseudoscopic lens focusing it! A focused-image hologram is a hologram of the lens as well as of the image the lens produces.

You can make focused-image reflection holograms, too. The setup is similar to other types of reflection holograms (except, of course, for the presence of a lens); and if you are using a single object-illuminating beam it is again easier to arrange the setup on its side. As the image is inverted for viewing, you will have to 'invert' the reference beam, so that it comes from 'above' with reference to the focused 'object' and not the real object. However, if you intend displaying the hologram reversed, the reference beam should come from 'below' with respect to the focused-image 'object'.

Focused-image holograms, as already mentioned, will reconstruct in white light. Furthermore, they do not even demand high spatial coherence. You can use an ordinary lamp-bulb, and even, under some circumstances, a fluorescent tube. In this respect they resemble other types of image-plane hologram such as the open-aperture holograms described later. If you want an idea of what a focused-image hologram image looks like, and you have a 35 mm camera with a removable lens, take it off the camera and hold it steadily about 10 cm (4 in) from an object. Focus your eyes in front of the lens so that you see the inverted real image. Watching this image, move your head from side to side. What you see is precisely what you will see when you view the hologram, except that the image is inverted.

Focused-image holography works best for shallow objects. The reflection type of hologram can make an attractive pendant or

94

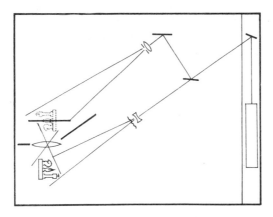

Focused-image reflection hologram for real-image viewing. This gives a bizarre effect on reconstruction, as the image of the lens is itself pseudoscopic and colour-fringed.

ornament, which is effective in ordinary light owing to the lax viewing requirements. The object, in general, should be smaller than the diameter of the lens aperture; you can use a bigger object if you use a pair of condenser lenses from a photographic enlarger. At the end of this chapter there are instructions on how to make a 300 mm (12 in) square liquid-filled lens from acrylic sheet. The technique also lends itself to a certain amount of artistic creativity. If you use a flexible Fresnel lens (sold by educational stationers as a 'magnifying sheet'), you can produce some interesting distortions of the image. You can also stretch or compress the image in one direction, by using two cylindrical lenses of differing focal lengths with their axes at right angles, like the anamorphic lenses employed for wide-screen cine projection. You can use a multi-lens surface acrylic sheet close to the film to give a multi-image 'fly's eye' effect. You can also combine a normal hologram of an object with a focused-image hologram by setting up a small hand lens so that it produces an image in the film plane, while the original object is still visible behind it.

Single-beam holograms

It is a simple matter to produce single-beam holograms using a sandbox. With the objects set sideways and the beam coming from 'above' you can get dramatic lighting effects with transparent objects. You can also make transmission holograms of opaque objects using a single beam, but it is advisable to soften the strong backlighting with a reflector close to the object.

Single-beam transmission holograms have great depth of field, up to 50 cm (20 in), and using this technique you can get some very striking parallax effects with long objects.

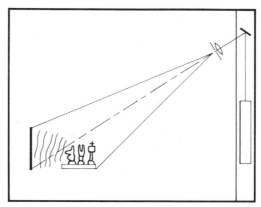

Single-beam transmission hologram.

Single-beam reflection holograms are also very easy to make with a sandbox. With the long throw available you can make these up to 8 × 10 in, and use either the standard virtual-image method or the pseudo-object technique for 'orthoscopic' real images. The depth of field is somewhat restricted: 50 mm (2 in) is about the limit.

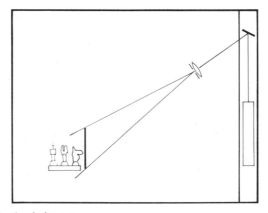

Single-beam reflection hologram

Real-image transmission holograms

The spectacular real-image holograms which caused such a sensation some years ago, at their first public appearance, are not

96

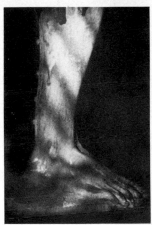

Holography and the preservation of statuary.
Above Ralph Wuerker and Giorgio Guattari in the restoration laboratory, San Gregorio, Venice, preparing to make a hologram of Donatello's wood-carving of St John the Baptist, using a Q-switched ruby laser.
Below, *left* Head of the statue, showing face deterioration.
Below, *right* Right leg of the statue. By warming the leg between two exposures, an interferogram is obtained. The irregularities in the fringes indicate hidden cracks and patches.

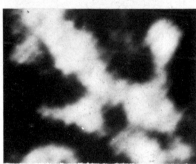
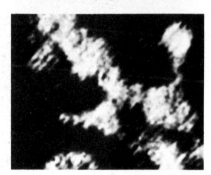

Fourier-transform holography and image deblurring.
Top Reconstruction of a Fourier-transform hologram. Two real images are produced, one of which is inverted. The bright spot in the centre is the zero-order beam.
Centre Deblurring by deconvolution. The deconvolution filter was a Fourier-transform hologram of the blur function, in this case a horizontal line 1·5 mm long.
Bottom Improving the image quality of an electron micrograph of a virus particle. The convolution filtering results in considerably improved resolution.
Courtesy of George W. Stroke, Stony Brook University.

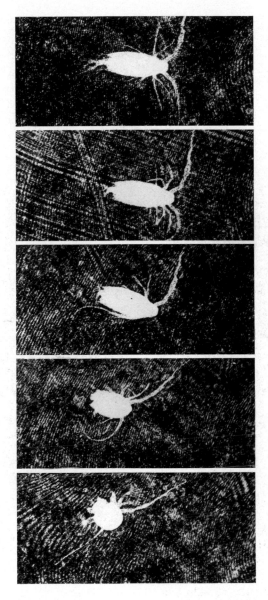

Holographic ciné-micrography. This sequence by L. O. Heflinger shows a copepod swimming. The creature is about 5 mm in length. The frames are focused-image holograms on 35 mm film, exposed to a pulsed argon or xenon laser at 50 per sec. The reconstructed images can be examined frame by frame through a microscrope. *Photograph by courtesy of Ralph Wuerker, TRW Systems Inc.*

'Head of Aphrodite' (1978), an achromatic white-light transmission hologram by Stephen and Jeanne Benton. This sculpture is housed in the Boston (Massachusetts) Museum of Fine Arts. This is a stereoscopic pair. *Photograph by Itsuo Sakane.*

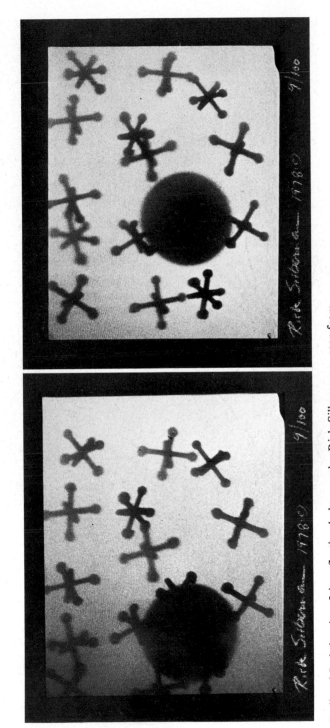

'Ball and Jacks', a 4 × 5 in reflection hologram by Rick Silbermann, seen from two viewpoints. *Photographs by Linda Law.*

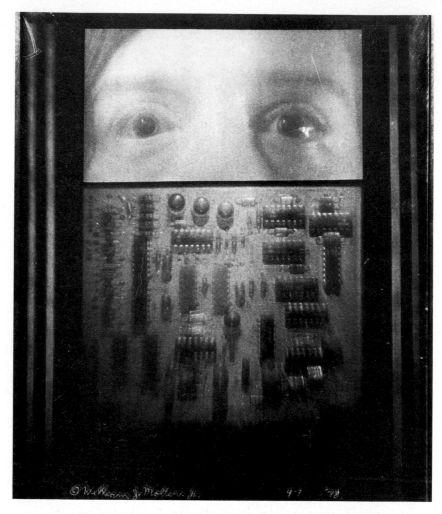

One way of turning a transmission hologram into a reflection hologram is to back it with a mirror, as in this example by Bill Molteni. It combines a pulsed-laser portrait with a continuous-wave laser still-life, in one hologram. *Photograph by Linda Law.*

Above A double-exposure hologram by Margaret Benyon, showing a bird in a box which is seen as a continuous surface. The viewer sees the front and the back of the box simultaneously.

Below This laser transmission hologram by Nick Phillips produces an image in front of the plate which is so realistic that people have thought the telephone is actually there. In this photograph, the hand is real. *Photograph by Theo Bergström.*

'Centerbeam' (1977–78), an outdoor holography display centre designed at the Massachusetts Institute of Technology. During the day it tracks the sun and at night the holograms are illuminated by artificial light. *Photograph by Palumbo*.

particularly difficult to make using sandbox techniques. The principle is essentially simple: you use the pseudoscopic real image of a transmission hologram as the object for a second hologram. The real image from this second hologram has had its parallax reversed twice, and is therefore orthoscopic.

In order to obtain a satisfactory image it is necessary to collimate the reference and reconstruction beams for the first (or master) hologram; this is the main problem. If you can get hold of a pair of old condenser lenses from a whole-plate or 8 × 10 in enlarger, one of them will do fairly well, though the focal length will probably be rather short, necessitating the use of a ×60 microscope objective as a beamspreader. You may be able to find an old government-surplus 36 in *f*6.3 aerial reconnaissance camera lens (there are still some around), and this will give very good collimation. A reflector telescope objective also makes an excellent collimator. For a 4 × 5 in hologram it needs to be about 90 cm (36 in) focal length and about 15 cm (6 in) in diameter. You can buy kits for grinding your own mirror from a blank, and this will save about two-thirds of the cost. You can get blanks up to 12½ in in diameter, and this size will cover an 8 × 10 in film. You can also make your own liquid-filled lens using the instructions on page 102.

The most elegant way of producing a collimated beam is by holography. Stephen Benton has devised a method of producing a holographic collimator, and this is described in Appendix 4.

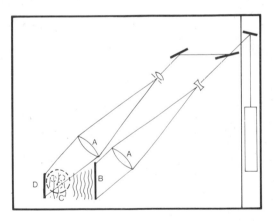

Real-image hologram for monochromatic reconstruction (both beams collimated).
A Collimating lenses.   **B** Master hologram.   **C** Pseudoscopic real image.
D Film for final hologram. Note that the reference beam should be slightly converging.

A good-quality unbleached transmission hologram will make a satisfactory master. It should be as large as possible, preferably twice the size of the final hologram, otherwise the 'window' will be somewhat restricted. Theoretically, both beams should be collimated. If the final hologram is to be viewed in a diverging reconstruction beam, its reference beam should be slightly converging. In practice, as we shall see, the requirements are not as stringent as this, provided we accept some limitations on such factors as depth of field.

Open-aperture holograms

If you are prepared to accept restrictions on the image space, you can produce a final hologram which does not need coherent light

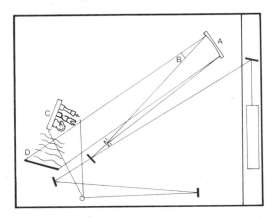

Master hologram using telescope objective as collimator. **A** Objective. **B** Angle between incident and reflected beams must be kept small. **C** For open-aperture holograms object should be on its side and needs to be shallow in depth. **D** Angle of incidence of reference beam should be at least 30°.

for viewing. It is called an open-aperture hologram, to distinguish it from a slit-aperture hologram (see below). The object should be shallow, preferably not more than 13 mm ($\frac{1}{2}$ in) deep, and approximately 125 mm (5 in) from the film which is to become the master. Set up the object on its side, with the reference beam (collimated) striking the film from 'below' the object at an angle of incidence of at least 30°. Process the master in the conventional way; that is, fixed and not bleached. To make the final hologram, set up the master, again with a collimated beam, to produce the pseudoscopic real image. Find the real image with a piece of white card, and move the card away from the master until the image just begins to be blurred. This is the position for the filmholder. Now set up your reference beam for the final hologram. This does not need to be collimated. The beam

98

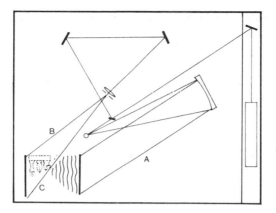

Final open aperture hologram with telescope objective collimator. **A** Collimated reconstruction beam. **B** Uncollimated reference beam at least 45° to final hologram film. **C** Pseudoscopic image acting as object. This should straddle, or be close to, film plane. Note that the depth of field is much less than shown here.

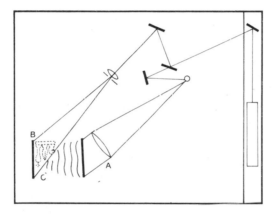

Final open-aperture hologram with lens collimator (Indications as in telescope-objective version.)

intensity ratio for the final hologram should be rather higher than usual, between 5:1 and 10:1. It is possible to get good holograms with ratios as high as 80:1.

You can examine open-aperture holograms by sunlight, or by the light of a projector lamp. They are easy to make, and unlike reflection holograms, do not require the finest-grain holographic film: you can use Agfa-Gevaert 10E75 or Kodak S0253 emulsions, which allow a much shorter exposure.

White-light transmission holograms

This type of hologram, sometimes called a Benton hologram after its inventor, is a transmission hologram which can be viewed in white light and has better definition, and a much greater depth of field, than open-aperture holograms. In order to achieve this the vertical element of parallax is sacrificed. The fringes which normally provide vertical parallax are replaced by an optically generated diffraction grating which disperses the white light into a spectrum in the vertical plane. This means that at any particular elevation the eye receives only a narrow band of wavelengths. By changing the elevation the image goes in succession through all the spectral hues from red at the top to violet at the bottom – hence the name 'rainbow hologram'. It would be more appropriate, however, to call it a 'slit hologram' to distinguish it from an open-aperture hologram.

As with real-image and open-aperture holograms, making a white-light transmission hologram is a two-stage process. The

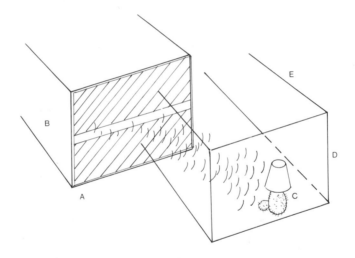

Principle of white light transmission (rainbow) hologram.   **A**   Master hologram masked to a horizontal slit.   **B** Collimated reconstruction beam from above produces **C** Pseudoscopic real image close to plane of **D**, final hologram film.   **E** Reference beam from 'below' (real image is projected inverted).

first stage is to make a conventional amplitude transmission hologram, using a collimated reference beam coming from above. This master hologram is used to produce a pseudoscopic real image which acts as the object for the final hologram, as in open-aperture holography. However, in this case, instead of illuminating the whole hologram we illuminate only a narrow horizontal band of it, thus losing the vertical parallax information. To view the result you set up the final hologram with the

100

direction of the reconstruction beam reversed. You are, in effect, viewing the holographic image through a real image of the slit; if you were to reconstruct the image with laser light, you would be able to see it only with your eye directly aligned with the 'slit'. However, the white reconstruction light forms an image of the slit in a different horizontal plane for every wavelength. The result is that if you look at the hologram from a high viewpoint you see a red image, and if you look at it from a low viewpoint you see a violet image. In between, the image takes all the hues of the spectrum in turn.

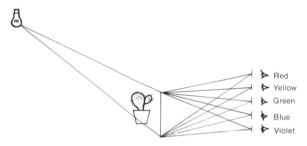

Reconstructing a white-light transmission hologram. The colour of the image depends on the height of the viewpoint.

To obtain the best-quality holograms the object needs to be rather farther away from the master film than normal, about 30 cm (12 in). This means that in order to get sufficient parallax you need a film at least 25 cm (10 in) wide. However, it is not necessary to use a whole sheet of 8 × 10 in film, as you are going to use only a narrow strip of it to produce the image for the second hologram. In fact, you can produce a perfectly satisfactory master from a length of 35 mm or 70 mm holographic film.

Making a collimating lens

You will probably also be relieved to know that you do not need to hunt for a 12 in diameter collimating lens. As only a narrow strip of the master hologram is illuminated, there is no point in collimating the reference beam in a vertical direction. This means that you can use a cylindrical lens, and you can easily make one of these from acrylic sheet. A plano-convex shape is both the simplest and the best. Use 3 mm thick sheet for the flat side and 1.5 mm for the convex side. The radius of curvature is approximately half the focal length, which should be equal to the distance between the lens and the beamspreader. For a ×40

101

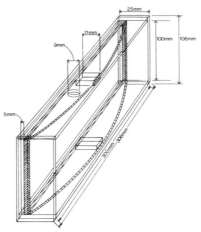

Construction of a cylindrical acrylic lens. Curved surface is 1.5 mm thick, remainder 3 mm. Lens is filled with liquid paraffin or glycerol.

microscope objective and a master 25 cm wide you will need about 2 m focal length, giving a radius of curvature for the convex side of about 1 m. If the lens is 30 cm (12 in) wide, the separation at the centre line will be 11 mm (0.44 in). The convex sheet requires to be 1 mm (0.04 in) longer than the flat sheet. The height of the lens is immaterial: 75–100 mm (3–4 in) is satisfactory. You also require two endpieces 25 × 325 mm (1 × 13 in), cut from 3 mm sheet. Get the sheets cut to size when you buy them, as it is difficult to cut acrylic sheet with an ordinary saw.

First, bore a hole in one of the endpieces near (but not at) the centre, using a 9 mm ($\frac{3}{8}$ in) drill. This is so that you can fill up the completed lens with liquid. Next, fix the flat sheet to the endpieces using acrylic cement. You can buy this ready made, or you can make your own more cheaply by dissolving scraps of acrylic sheet in chloroform or acetone, to the consistency of syrup. Give several coats, allowing each one to dry before applying the next, so that the joint is liquid tight. Now cut two separators 11 mm wide and fix them to the centre top and bottom. Add the curved sheet, holding it in place until the cement is completely dry.

The last task is to fill the lens with glycerol or liquid paraffin. Do this at room temperature, and leave a bubble to allow for expansion and contraction. Plug the hole with a synthetic rubber bung. Your lens is now ready for use. A word of warning, though: although acrylic sheet is optically flat, it is much softer than glass, and the surface is easily damaged. Keep your lens wrapped in tissue when you are not using it, and treat it as gently as you treat your front-surface mirrors.

102

Practical layouts

White-light transmission holograms are usually viewed with the reconstruction light coming from above; so, as is usual for this arrangement, it is convenient to mount the object on its side. The strip of film will then be vertical, as will be the collimating lens. The reference beam should strike the film at an angle of incidence

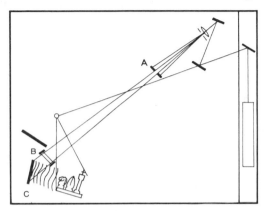

Layout for white-light transmission master hologram. **A** Light baffles. **B** Cylindrical lens (set upright). **C** Film at 30° incidence, set vertical.

of at least 30° from 'above'. The beam intensity ratio should be between 4:1 and 8:1, the lower figure being preferable. Develop and fix the film without bleaching.

For the second stage, you need to turn the master round and illuminate it from the other side. You still require the collimating lens; but you want the reconstruction beam spread in only one direction. You therefore need a cylindrical lens or mirror. You can make a good lens from a 12 mm (0.5 in) diameter test tube filled with liquid paraffin or glycerol, or from a glass rod of the same diameter but a cylindrical mirror made from a 25 mm (1 in) diameter roller bearing is simpler. You should be able to get one of these where you got your ballbearings. This will project a line of light on the master hologram. The breadth of this line is quite important, as a broad line results in poor definition and a narrow line results in increased speckle. The optimum is 3–6 mm. As the divergence of an unspread laser beam is about 1 mm per metre, you will need to use the full length of the sandbox. Mask off the unwanted part of the master with opaque tape, and set it up in the beam vertically, as you did before, aligning the slit carefully with the beam.

Examine the real image using a white screen and check that it is the right way up; that is, the reference beam for the final

103

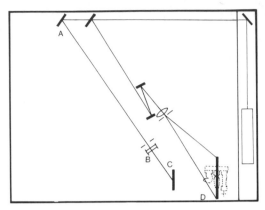

Layout for final white-light transmission hologram. **A** Roller bearing acting as one-dimensional beam spreader. **B** Cylindrical collimating lens with limiting slit. **C** Master hologram. **D** Pseudoscopic real image and final hologram film.

hologram comes from 'below' the image. The angle of incidence of the reference beam should be at least 45° and the beam intensity ratio about 5:1, though ratios up to 10:1 will still give satisfactory results. Benton's own recommendation is a reference beam angle of incidence of 54.5° and a reconstruction beam angle of 45°, which gives a greenish image. If the master and final hologram are separated by about 20 cm (8 in) the final image will be virtual; and if the separation is about 30 cm (12 in) it will be real. At about 25 cm (10 in) it will straddle the film plane. In the last case the colour will appear much less saturated, and the image will reconstruct with a white-light source that is not spatially coherent. A lamp with a vertical straight filament will give an almost achromatic image. If you route the reference beam for the final hologram to the rear surface of the film, you will obtain a reflection rainbow hologram, and this will also reconstruct in white light.

If you are prepared to go to the trouble of making a second cylindrical lens, you can produce speckle-free deep-image holograms that are also sharp. The technique was devised by Emmett Leith and Hsuan Chen. The lens is situated immediately behind the master hologram, and you can use it as a support for this. The curvature is along the shorter dimension, i.e. across the slit. It should have a focal length such that it focuses the image (in the horizontal sense) approximately in the plane of the film. For the setup described here this should be about 150 mm (6 in). The method of construction is exactly the same as for the collimating lens, except for the direction of curvature. If you make the shorter dimension of the lens 75 mm (3 in) the separation of the inside surfaces should be 10 mm (0.4 in). Acrylic sheet is too rigid to give the curvature required for the curved surface of this

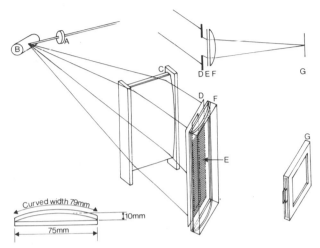

Minimizing blur and speckle in deep-image holograms. **A** 200 mm concave lens. **B** Roller bearing. **C** Cylindrical collimating lens. **D** Restricting aperture (max. 29 mm). **E** Master hologram. **F** Cylindrical lens focusing image in 'horizontal' plane on **G**. Final hologram film.
Note: Arrangement is on its side, so that 'horizontal' is in fact vertical in the set up.

lens, so acetate sheet should be used instead. When you have finished the lens, fill it with glycerol or liquid paraffin as you did with the collimating lens. The curvature of the surface of this lens is rather higher than for the collimating lens, and you need to bend the acetate sheet in several stages over a period of a day or so, before fixing it permanently. It is also a good idea to use two side-strips, for reinforcement, so that all sides are boxed in (as in the lens on page 106). Use a carpenter's vice to hold the lens while the cement is setting.

With this technique you can employ the full usable width (24 mm) of 35 mm film. With 70 mm film you can go up to the maximum useful width of 29 mm ($1\frac{1}{8}$ in). In order to broaden the beam sufficiently across its narrow dimension, you will need to spread it slightly by means of a 200 mm concave lens set close to the roller bearing.

Building a 'spherical' collimating lens

Being cylindrical, the lens described on pages 101–3 collimates the light in one direction only. If you were to put two of these lenses together, back to back, the axis of one being at right angles to that of

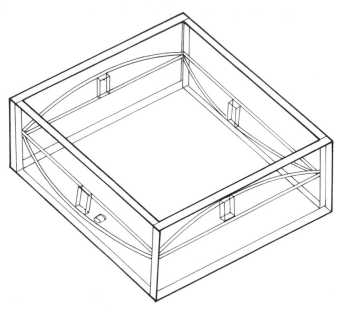

A homemade liquid-filled lens. Build the outer box first, and insert the lens surfaces only when the cement has thoroughly hardened. This lens has dimensions suitable for focused-image holograms, or for collimating a beam spread by a ballbearing of diameter 6 to 8 mm. Dimensions of curved surfaces for longer focal lengths are given in the text.

the other, you would be able to collimate light in both horizontal and vertical planes. You would, in fact, have produced what amounts to a spherical lens, though with cylindrical surfaces only. Such a lens will enable you to make large open-aperture holograms, or to produce a collimated or converging reference beam for the final hologram with the slit technique.

This lens is no more difficult to make than the one described on pages 101–3. The first step is to make a cylindrical lens by this method, using 6 mm acrylic sheet 300 mm (12 in) square for the plane surface and 1.5 mm sheet 300 × 301.2 mm ($12 \times 12\frac{1}{10}$ in) for the curved surface. The endpieces should be 44 × 300 mm ($1\frac{3}{4} \times 12$ in), and you will need two sidepieces 44 × 312 mm ($1\frac{3}{4} \times 12\frac{1}{2}$ in) in addition. Build the lens so that the plane side is on the centre line of the endpieces, using spacers 11 mm (0.44 in) wide. Now fix the two sidepieces in position and add a further curved surface of the same dimensions as the first, but oriented at right angles to the first and on the other side of the plane surface. Bore a hole in one endpiece and one sidepiece, and fill the cavities with glycerol or liquid paraffin.

The plane surface is not essential to the lens: in fact, optically it would be better to dispense with it. If you make the outer frame

106

first, holding it in place with stout rubber bands until the cement is completely dry, you can fit the two curved surfaces without the necessity for the flat sheet. As before, the width of the four spacers should be 11 mm.

This lens has a focal length of approximately 2 m. If you require a focal length of 1 m, use sheets size 300 × 302.5 mm (12 × 12.05 in), and spacers 22 mm (0.88 in) wide. You can make lenses of shorter focal length for focused-image holograms, and, of course, the curvature of the surfaces is much greater, and they will require prebending for 24 hr before fixing. The dimensions for this lens are shown in the diagram and by making the surfaces depart from circularity, or by using two surfaces of differing curvature, you can introduce some intriguing forms of distortion. You will need approximately 9 litres of liquid to fill the lens shown. Use liquid paraffin or mineral oil, which are much cheaper than glycerol. You can also use water that has been boiled, but the focal length will then be 50 per cent greater.

Summary of Do's and Don'ts in holography

DO:
Wait until evening before you start work: there will be less stray light, and much less vibration.
Site your sandbox well away from radiators, convectors, refrigerators and water-pumps, and if possible on a concrete floor.
Carry out an interferometer test before each holography session.
Heap up sand round any elevated components.
Ensure any asymmetrical optical components are properly counterbalanced.
Allow at least 15 min for laser to warm up and optical components to settle before starting work. This particularly applies to reflection holography.
Check that the speckle pattern on the screen is absolutely stationary before replacing it with the film.
Ensure all optical paths are equal to within about 1 cm before exposing.
Use the highest-resolution film (e.g. Agfa 8E75HD) for reflection holograms, and remove any antihalation backing with alcohol before using.

DON'T:
Move around during exposure.
Breathe heavily anywhere near the apparatus.

Cough, sneeze, talk or make any other loud noise during exposure.

Set up any component with the support tilted more than about 30°.

Leave exposed films for more than an hour or so before processing.

Look into an unspread laser beam, or allow a beam to fall where anyone unexpectedly entering the room may do so.

Allow anyone but yourself to operate a laser of more than 1 mW output power, or to enter the room where one is switched on, whether the beam is spread or not.

More advanced types of hologram

This a practical guide for the amateur, and up to this point we have given full instructions on all optical layouts. This chapter does not do so. Some of the techniques described here are still under development, and some require special equipment or optical laboratory conditions. Nevertheless, many of these newer techniques are amenable to sandbox methods; by now you will no doubt be able to set up a holographic arrangement from a schematic diagram. You may well be able to contribute something of value yourself.

Holograms on exhibition

In exhibitions of holography it is often the sheer size of the holograms that makes them so spectacular. Most of them are simply conventional holograms on a very large scale, made with powerful argon–ion lasers operating in the green and blue regions of the spectrum. The most dramatic kind is probably that in which a real image is projected into the space in front of the hologram. In 1972 Cartier, the New York jewellers, displayed a projection hologram in which a hand wearing a diamond ring and holding a necklace was projected right out into the street, to the astonishment of passers-by. One of them is said to have shouted that it was the work of the devil, and to have attacked the image with her umbrella.

You have seen that small real-image holograms, even those with white-light reconstruction, are not difficult for the amateur. It is only the scale of operations that makes things complicated for these giant holograms. Once you go above 8 × 10 in a few headaches appear; for example:

1 you need plates rather than film, and the tolerances in setting up are small: and as light losses are high, you need a powerful laser;
2 steadiness requirements are very stringent indeed;
3 any problems of resolution in the master hologram are compounded with those in the final hologram;
4 in order for the final hologram to give a bright enough image for realism, it must be a phase hologram; to produce one with the required resolution imposes severe constraints on processing technique.

New processing methods, as described in Appendix 3, have made the last requirement less daunting. Nevertheless, to make really large holograms you need three things: endless patience; a taste for utter precision; and a sponsor with a fairly bottomless pocket. If you have all three, the world of large holograms is at your feet.

Pulsed-laser holograms

The very first lasers were pulsed, with a duration of about 1 millisecond (ms). You may wonder why they were not used at the time for making holograms of moving or even living subjects. The main reason was poor coherence length, caused by the complicated structure of the pulse itself, which was not a single pulse but many small ones. This problem was solved by what is called *Q-switching*. A cell of an opaque dye, which bleaches instantaneously when the light intensity reaches a certain level, is inserted in the optical cavity. This confines the light until it has built up sufficiently, then lets it go in a single giant pulse. We can amplify this pulse (which typically lasts only a few tens of nanoseconds) by passing it through a second laser cavity. We can limit the spread of frequencies by inserting a device called a *Fabry-Pérot etalon* into the optical cavity. This is itself a small optical cavity which resonates at only one visible frequency, corresponding to a wavelength that is an integral submultiple of the cavity width. Its operation is analogous to the way the walls of an alley produce an echo of your footsteps at a definite pitch, which depends on the width of the alley, rejecting all other frequencies that may have been present in the original sound. *Q* is a symbol used in electronics engineering to describe the narrowness of a bandwidth of frequencies. Many holograms of

110

living people have been made: one of the earliest successful holographic portraits was of the late Professor Dennis Gabor himself. More recently, pulsed lasers have been used to make holograms of many types of moving objects, from fruit-flies to bullets, as well as ciné-holograms.

The main obstacles for the amateur are the sheer cost of etalons, dye switches and the laser itself, as well as the risks to eyesight from megawatt pulses of light. In the United States and several other countries it is necessary to hold a licence in order to operate a powerful laser. However, there is no inherent difference between holography with a continuous-wave laser and with a pulsed laser except, of course, that in the latter case movement of the subject is no problem. It is possible to hire lasers, and even to rent entire holographic studios, but it is still an expensive business.

Holographic interferograms

These are used mainly for scientific purposes, and some of their uses are shown in the Plates. They are of two types. The first is a double-exposure technique. If you set up and expose a holographic film, then without disturbing any of the optical components you make a second exposure, any change in the shape of the object during the intervening time will appear in the developed hologram as a set of contour fringes, one for every wavelength of movement. This technique also makes it possible to visualize shock waves in air if the first exposure is of the undisturbed air and the second while the object is in transit. The second type of holographic interferogram is made with a single, longer exposure, and is used to produce interferograms of steadily vibrating objects such as loudspeakers. This speaker cone moves between two extremes, and at these extremes it is momentarily stationary. The positions between the extremes do not record because of the movement, but the extremes themselves do, giving what is effectively a double exposure which shows the pattern of vibration of the cone contoured by interference fringes. The shock-wave type is confined to sophisticated laboratories, of course: however, double-pulse lasers are available, always provided you can afford one! The vibrating-object type of interferogram is simple enough in principle, and requires no special apparatus; but you will probably have to go to some trouble to ensure none of the vibrations can be transmitted to your optical equipment. It is easy to make straightforward interferograms showing distortions. For example, you can warm your object with a hair-dryer for a few seconds between exposures, or use an object such as a plastic phial held in a vice, which you tighten a fraction of a turn between the exposures. Remember, though, that any movement of more than a few wavelengths will cause the fringes to be so close together as to be invisible.

Volume and multiple-image holograms

In a so-called *plane hologram*, the object and reference beams fall on the film at approximately equal angles of incidence. The emulsion is thin, and the interference fringes are formed perpendicular to the emulsion surface. However, if the object beam is at zero incidence (i.e. perpendicular to the surface) and the emulsion is comparatively thick, the fringes will lie at an angle to the surface, and will be distributed through the thickness of the emulsion, forming what is called a *volume hologram*. The reconstruction of the object wavefront is achieved in a way somewhat

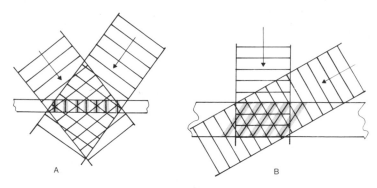

'Plane' and volume holograms. **A** In a so-called plane hologram the emulsion is thin. Object and reference beams are at approximately the same angle of incidence and the fringes are perpendicular to the surface. **B** In a volume hologram the emulsion is thick. Object and reference beams are at differing angles of incidence and the fringes are oblique to the surface.

different from that of a plane hologram. The principle is known as *Bragg diffraction* (after Sir William Bragg and his son Sir Lawrence Bragg, who first described the phenomenon), and, like the reconstruction of the object wavefront in a reflection hologram, it depends on constructive interference.

Volume holograms have a high *diffraction efficiency* (that is, the zero-order beam is suppressed), especially if bleached. The main difficulty is in the processing. If there is any shrinkage or distortion the diffracted waves will not meet the Bragg condition, and the result will be a weak, degraded image. Experimental

112

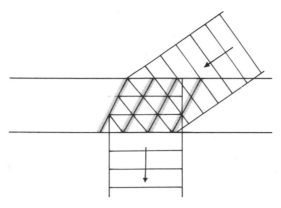

Reconstruction of a volume hologram. The reflected waves interfere constructively (Bragg diffraction).

emulsions up to 1 mm thick have been made for volume holograms intended for information storage, but there are severe problems in processing, and to date these have not been satisfactorily solved. However, even a 'thin' emulsion such as 8E75HD is still many wavelengths thick, and even so-called plane holograms have a volume element. The Bragg condition restricts the angle of incidence of the reconstruction beam to within a cone of about 20° angle; and as an amateur, you can exploit this directionality to make holograms containing more than one image.

First, make a holographic exposure of an object with the reference beam offset about 25–30° in one direction. Now make a second exposure of the same film on a different object, this time with the reference beam offset about the same amount in the other direction. When you reconstruct the image, a rotation of the hologram in the reconstruction beam will cause one image to dissolve into the other. You may be able to do this with three or even four images, depending on the film you are using.

An extreme version of this technique is a hologram which shows the front of an object when viewed from one side and the rear of it when viewed from the other. To make this convincing requires one of the image-plane techniques. A focused-image hologram is the simplest: you simply rotate both the object and the film through 180° between the two exposures. A better method is the open-aperture (pages 98–9) or the slit (pages 99–106) technique. You need to mount the object (on its side) in such a way that it can be rotated through 180° about a horizontal axis. Make two master holograms in succession, rotating the object through 180° between the exposures. For the final hologram, the image should straddle the film plane. Make an exposure using the first master,

113

then rotate the film about a horizontal axis so that the back of the film is turned towards the master. Now substitute the second master and make a second exposure. Each of the two exposures should be rather less than normal. A converging reference beam gives the best results, but is not strictly necessary. Mount the bleach-processed hologram using photofloods for reconstruction, one each side. You will be able to walk round the hologram and see both sides of the object: the impression of seeing it in the round is very powerful.

Achromatic holograms

The idea of a white-light hologram which reconstructs an uncoloured or *achromatic* image has long intrigued holographers. In 1977 Stephen Benton exhibited a successful achromatic hologram of a Greek sculpture, a head of Aphrodite, and demonstrated several methods of achieving an achromatic image. Of course, any image-plane hologram with a shallow image can be reconstructed using white light that is spatially not very coherent, and this will give an image that is not strongly coloured; but for deeper images there is a problem. In the rainbow-type white-light transmission hologram each wavelength reconstructs the image (and the real image of the slit) at differing magnifications, and at differing distances from the hologram. The red is closest and smallest, and the blue farthest away and largest. The line of slit images is tilted at an angle of about 35° to the hologram. In order to produce an achromatic image it is necessary to produce a number of spectra, and these need to be aligned so that at each point there are red, green and blue slit images which precisely coincide.

A crude method which goes some way towards achieving this is to use a vertical-filament lamp for reconstruction. However, this does not dispose of the problem of magnification differences, and can result in unacceptable loss of resolution. By using three reference beams at slightly differing angles of incidence and collimated to differing convergences the red, green and blue wavelengths can be made to reconstruct slits and images of the same size and coincident in space. An alternative approach, the one used for the Aphrodite hologram, is to use multiple beams to illuminate the master hologram. A 'diffractor plate', made holographically, produces the effect of three point sources of light along the tip angle of the spectrum, and this is used to produce the three reconstructing beams for the master hologram image when making the second hologram. As might be expected, this

114

second hologram is badly blurred; but if it is set up to produce a real image for recording on a third plate (with a collimated beam illuminating the second master) this final hologram contains the images of all the slits. The spectra coincide in space, so that the image is achromatic over a wide angle of viewing. Under the auspices of Stephen Benton, courses are currently being offered in both achromatic holography and the making of collimators by holographic means (see Appendix 4). Benton himself offers courses in advanced holography roughly twice a year at Polaroid Corporation, Cambridge, Massachusetts, USA.

Emmett Leith and his associates have adopted a different approach. The principle is basically fairly simple. It also uses the rainbow principle, with a 10 mm wide slit; and the master hologram image is focused in the vertical plane by a cylindrical

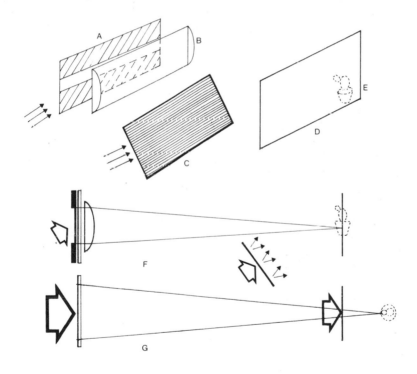

White-light transmission hologram with one-dimensional diffuser.   **A**   Master hologram masked to a 10 mm slit.   **B**   Cylindrical lens focuses image in film plane (verticals only).   **C**   One-dimensional diffuser scatters reference beam in vertical plane only.   **D**   Holographic film.   **E**   Pseudoscopic real image from master hologram.   **F**   In the vertical plane the configuration is an image–plane hologram.   **G**   In the horizontal plane it is an in-line hologram. Both types will reconstruct with either coherent or incoherent light.

lens of the type on page 105. The novelty is in the reference beam; it passes through a diffuser plate which diffuses the beam only in the vertical direction. This arrangement, in fact, is another way of producing a multiple-reference beam. It gives an achromatic image when viewed by white light and, unlike the usual slit hologram, it can be viewed by monochromatic light over a wide vertical angle. The image shows some *astigmatism* (that is, the vertical aspects of the object are not focused in the same plane as the horizontal aspects), but this is not noticeable.

The best method of making the diffuser is to score a piece of acrylic sheet with sandpaper. As there must be no variation in the horizontal direction you will need to set up some kind of jig, and, to be on the safe side, use a fresh sheet of sandpaper for each pass. The number of passes you need depends on the amount of pressure applied and the grade of sandpaper (No. 0 is probably best). Do not overdo the diffusion. The beam angle is critical in this method: about 35° incidence is approximately correct.

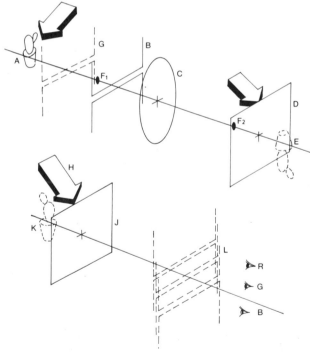

One-step white-light transmission hologram (pseudoscopic image). **A** Object. **B** Slit. **C** Imaging lens. **D** Film. **E** Real (optical) image. **F_1** and **F_2** Principal foci of lens. **G** Virtual image of slit. **H** Reconstruction beam (white). **J** Hologram (= **D** flipped over top to bottom). **K** Pseudoscopic image. **L** Real (rainbow) image of slit.

One-step white-light transmission holograms

By using a slit in conjunction with a focused-image holographic arrangement, it is possible to produce white-light transmission holograms in a single step. The principle is quite simple. One method produces a pseudoscopic image. The second method produces an orthoscopic image, though parallax is severely restricted unless you use a lens with a large aperture, as suggested on page 93.

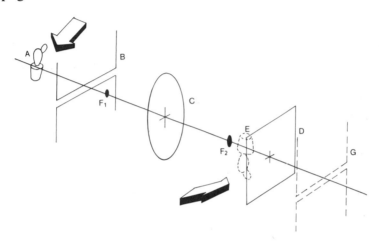

Arrangement for orthoscopic image. **A** Object. **B** Slit. **C** Lens. **D** Film.
E Real image of object. **F**₁ and **F**₂ Principal foci of lens. **G** Real image of slit.
The developed hologram is viewed 'right way round'.

P. Hariharan has successfully disposed of the problem of restricted field of view by employing a large concave mirror instead of a lens. The mirror has a radius of curvature of 550 mm and a diameter of 600 mm, giving a horizontal field of view of 50°. He uses a 60 mm slit between the object and the mirror, which reduces the effect of aberrations and greatly increases the brightness of the image. As the (real) image of the slit is situated 2 m in front of the hologram and is magnified $6\frac{1}{4}$ times, there is some 10° of vertical parallax.

He has extended this technique to produce what would seem to be the ultimate in hologram realism: a white-light one-step orthoscopic image-plane reflection hologram in colour, which reconstructs with a car headlamp bulb in ordinary room lighting. The system requires two lasers, as do all natural-colour holograms; but the mirror between them is dichroic; that is, it reflects red light and transmits blue and green, so that it does not need to be removed between exposures. He shows that the use of

117

two plates (one for the red and one for the blue and green images), actually gives better and brighter images than could be obtained on a single plate, however good, because each hologram produces less individual scatter. In reconstruction, the 'red' plate with its stronger image is mounted behind the weaker-image 'green and blue' plate. The plates are fixed together with optical

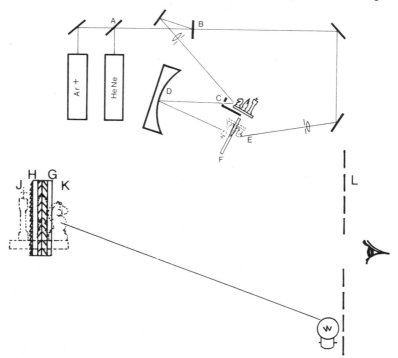

Natural colour white light reflection hologram (Hariharan's method). **A** Dichroic mirror reflects red, transmits blue and green. **B** Variable–reflectance beamsplitter. **C** Limiting aperture. **D** Concave mirror. **E** Real image (optical). **F** Holographic plate. **G** 'Blue and green' hologram. **H** 'Red' hologram. **J** Black acrylic paint. **K** Orthoscopic image in plane of hologram. **L** Real image of slit limits vertical parallax to 10°.

cement, and the rear surface of the glass is sprayed with a flat black acrylic paint.

Integral holograms

This type of hologram, sometimes called (more accurately, if a little long-windedly) a cylindrical holographic stereogram and also known as a multiplex hologram, is an intriguing hybrid of photography and holography. The image, which may be larger or smaller than life size, appears to be within a transparent cylinder

118

(the hologram), and with full 360° parallax. The cylinder may be
driven so that it rotates slowly, and the image then rotates with it.
Alternatively, where appropriate, only a 120° portion of a
cylinder is used, and the mounting is fixed.

The integral hologram, developed by Lloyd Cross, starts life as
a more or less conventional ciné film. The subject is placed on a
rotating plinth, and the operator makes a ciné record of a 120° or
360° rotation. Alternatively, the camera moves in a circle round
the stationary subject. The resulting transparencies are used to
make a series of holograms in the form of adjacent narrow strips
on a single film, using the rainbow technique. The processed film
is mounted in a transparent cylinder and viewed using white light
from a short-filament lamp situated below the cylinder. As each
strip hologram represents a two-dimensional transparency, it
shows no parallax in itself; but when the viewer looks at the
virtual image within the cylinder, each eye looks through a

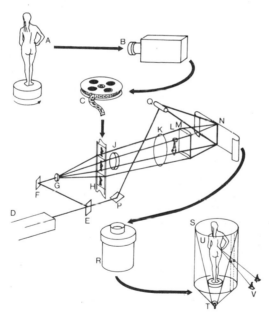

Making an integral hologram. **A** Model on slowly rotating plinth. **B** Cine camera
records 360° rotation. **C** Developed film. **D** Laser. **E** Beam splitter.
F Mirror. **G** Beam spreader. **H** Transparency. **J** Projector lens.
K Collimating lens. **L** Real image formed by projector lens. **M** Cylindrical
lens. **N** Film at focus of cylindrical lens. **P** Tilted mirror. **Q** Cylindrical mirror
spreads reference beam in one dimension. **R** Exposed film is bleach-processed.
S Developed hologram formed into 360° cylinder. **T** Reconstruction illumination
from below. **U** Virtual image of model. **V** Each eye sees image through a different
'strip', thus obtaining parallax stereoscopy. *Note:* the detail of the technique has been
somewhat simplified for clarity.

different strip, as in a conventional stereoscopic viewer. Thus the image as a whole has full horizontal parallax, though with slight jumps between the strips. In practice, the jumps are minimized by overlapping the holographic strips and by 'printing' each frame twice.

As the making of the holograms is a one-step process, the image is, strictly speaking, pseudoscopic; but this is irrelevant, as the transparencies are flat and without parallax. One advantage of this method is that as the original record is an ordinary movie film, there is no need to expose live subjects to high levels of laser light. They can also move during the recording, which adds the extra dimension of time to the display. If the cylinder is displayed rotating at the same speed as the original, any action will be 'replayed'. A well-known early example of this was 'Kiss II' by Lloyd Cross and Pam Brazier, a 120° fixed integral hologram depicting a girl (Brazier) winking and blowing a kiss.

As the original is a set of transparencies which can be made at any scale, an integral hologram, unlike conventional holograms,

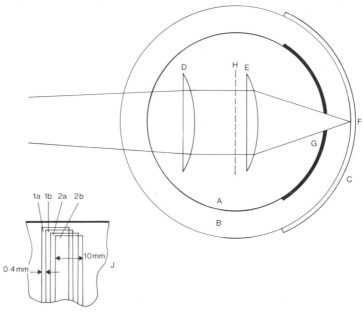

Layout of optical table for integral holograms. **A** Fixed table. **B** Outer part of table can be rotated. **C** 130° section of acrylic cylinder series as filmholder. **D** Collimating lens. **E** Cylindrical lens (both lenses may be liquid filled). **F** Focus of cylindrical lens. **G** Slit limiting width of exposed strip to about 10 mm. **H** Screen inserted temporarily to check position of optical image formed by projector lens. **J** Showing overlap of exposures (each frame is exposed twice). Rotation between exposures is one-sixth of a degree. The reference beam (not shown) strikes the film at **F** from above.

can show objects at any scale (usually much reduced). It can also depict outdoor scenes.

Optically, the setup for making integral holograms is somewhat similar to that described on page 116 for one-step slit holograms with pseudoscopic images. Making an integral hologram is a fairly formidable proposition for an amateur, as it requires the construction of an accurate jig, to hold the optics and the film. Commercial setups employ a large filmholder with a step-on system which advances the film by the right amount for each exposure, and can make the whole 360° on one strip of film. Rather than going to the trouble of making one of these, it is easier to make the film in three 120° strips. Allowing for a little overlap, the film size for a 40 cm (16 in) diameter acrylic cylinder will be 24 cm ($9\frac{1}{2}$ in) high, which is the standard height, by 48 cm (19 in) long. You can use a section of cylinder as a filmholder. There are three frames (six exposures) per degree, and this will need to be marked out on your table. The distance is just over 0.4 mm. The width of the image as focused by the cylindrical lens should be about 10 mm, and the reference beam (from above) should be the same width.

The total number of frames in a 360° rotation is 1080, which at 24 pictures per second represents one complete revolution in 45 sec. This needs to be exact. It is about three times as long as it would take to walk slowly round the cylinder. This means that in order to create a realistic movement the subject must move with exaggerated slowness. Rapid movement causes a peculiar kind of distortion when played back, as the eyes receive images that were not produced simultaneously.

As integral holograms are made on a continuous strip of film, they can be used as a somewhat limited form of ciné-holography. The result is displayed by being moved continuously across an aperture wide enough to permit simultaneous viewing by several people. The camera can also be held stationary, and though this loses the horizontal parallax the illusion of depth is still striking.

In common with other types of hologram involving cylindrical lenses, integral holograms show a certain amount of astigmatism. In particular, if examined from too close, the image appears elongated in a vertical direction. If you want to photograph the image in an integral hologram using a 35 mm camera, it advisable to use a lens with a focal length of at least 100 mm.

Holography in colour

There are two basic methods of producing an image in natural colours. The one used in photography is *subtractive*, that is, it uses white-light illumination, and the photographic image is in dyes which block all wavelengths other than the required ones. The method used for colour television, on the other hand, is *additive*: light of various wavelengths is mixed in the right proportions to give the required colour. Provided you choose the right bands of wavelength to achieve this you can produce every colour that exists; and you can produce very nearly every colour from mixtures of just three wavelengths, in the red, green and blue regions of the spectrum respectively. If you look closely at a colour television screen you will see that it is made up of tiny red, green and blue dots (or bands, on some tubes) which merge into continuous colours when you move farther away. The red (633 nm) beam of the helium–neon laser and the green (515 nm) and blue (488 nm) beams of the argon–ion laser can between them produce all visible colours except deep blue-violet and saturated yellow-greens. This is better than either colour television or photography can do. The helium–cadmium laser, operating at 442 nm and substituted for the blue argon wavelength, gives an even greater range; but the addition of a third laser complicates the optical setup.

As we have seen, there is no particular problem involved in making more than one hologram on the same piece of film. We can readily record three holograms of the same object with the same lighting arrangement, but using light from three sources: one red, one green and one blue. If the three beams are directed at the processed hologram, the image will be reconstructed in full colour.

Unhappily, things are seldom as simple as they seem. In colour holography, the main problem, at least with transmission holograms, is what is called *cross-talk*. Each beam reconstructs not only its own image, but also two spurious images from the interference patterns produced originally by the other two beams. These are of the wrong size, and in the wrong place, but they are near enough to get in the way. If the angle between the reference beam and the object beam is large the fringe spacing will be small, and the spurious images will be well separated from the genuine image. Unfortunately, this also leads to lower resolution. One successful method is to separate the reference beam by large angles. While this makes reconstruction more complicated, it does nevertheless work well.

By using a thick emulsion to produce a volume hologram, the spurious images are also reduced in intensity, as the Bragg condition is not satisfied for them. When bleached under condi-

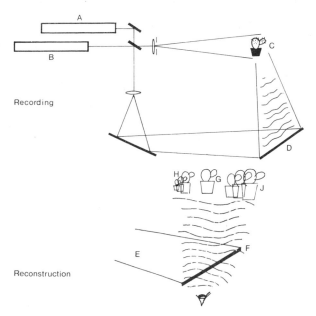

Three-colour transmission holography. **A** Helium-neon laser (red, 633 mm).
B Argon-ion laser (green, 514 nm and blue, 488 nm). *Note:* A better colour
reconstruction results if a third laser (helium–cadmium) is used for the blue
beam. **C** Object. **D** Film. **E** Three-colour reconstructions beam. **F** Hologram.
G Genuine image. **H** Three smaller spurious images (reconstructions of 'blue' and
'green' images by red light and 'blue' image by green light). **J** Three larger spurious
images (reconstructions of 'red' and 'green' images by blue light and 'red' image by green
light.

tions which avoid emulsion shrinkage, such holograms are fairly
satisfactory.

Transmission colour holograms of the conventional type have
found little favour outside specialized optics laboratories, as they
require accurately aligned triple reconstruction beams. A more
promising approach employs a multiplexing technique. By placing a graticule of alternating red, green and blue strips in contact
with the emulsion, three separate but interlaced records appear
on one film. The filter is left in place for viewing, which is by
spatially coherent white light. The graticule must be sufficiently
fine to be invisible when the hologram is viewed from the normal
distance. This method of reconstruction bears some resemblance
to the old Dufaycolor photographic process, which used a reseau
of tiny squares of red, green and blue for both taking and
viewing. It also avoids the 'cross-talk' problem as no spurious
images can be generated, but it requires a reconstruction beam of
high intensity.

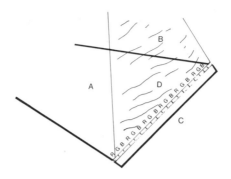

Multiplex colour transmission holography. · **A** Reference beam (three-colour). **B** Object beam (three-colour). **C** Film. **D** Filter consisting of narrow bands of red-, green- and blue-transmitting material in contact with film. This filter remains in place for viewing.

The problem of spurious images can be avoided by making a reflection hologram. This demands accurate alignment of the three beams only during the exposure, reconstruction being by white light. The spurious images are totally suppressed. The only problem – and it is a serious one – is that shrinkage of the emulsion during processing can cause the 'red' component to appear as green, the green to appear as blue, and the blue as invisible ultraviolet! However, recent research into new methods of bleach-processing has produced developer and bleach formulae which largely avoid this trouble (see Appendix 3). The remaining problem, the loss of resolution and diffraction efficiency due to the three images occupying the same space, is likely to be harder to crack, as the limitation is not merely practical but theoretical. The most satisfactory method at present seems to be

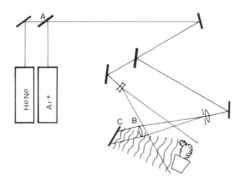

Making a three-colour rainbow hologram. **A** Removable mirror. **B** Collimating lens. **C** Master film. The final hologram is made with the usual optical layout, but it is necessary to make three exposures, one for each master, using the appropriate laser beams.

124

an adaptation of the white-light transmission technique, pioneered in Australia by P. Hariharan. This method produces a reconstructed image of high colour saturation, free from spurious images. As with other white-light transmission holograms, changing the height of the viewing point changes the hue of the image; so unless this phenomenon is to be exploited for effect, it is necessary to restrict the vertical angle of viewing by some sort of barrier.

Is it possible for the amateur to make natural-colour holograms? Yes, certainly. The only extra equipment you need is an extra mirror, some filters for balancing the three colours, and, of course, an argon–ion laser. Unfortunately, these lasers are very expensive, though you may be able to hire one.

At the time of writing there is no satisfactory holographic emulsion sensitive to all the visible spectrum. The emulsions made for helium–neon light are almost insensitive to green, and those made for argon–ion light are insensitive to red. The best way to tackle this problem is to use separate plates or films for the red and for the blue and green record. Thus the red record would be made on an 8E75 plate and the blue and green on an 8E56. One of these is exposed through the back, and the two plates are

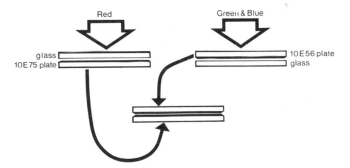

Arrangement of plates for final hologram. One of the two plates is exposed through the back. The two processed plates are bound up in register, emulsion to emulsion, for viewing.

then mounted in register for viewing. This method also goes part of the way towards avoiding the problem of loss of resolution caused by the presence of more than one image in the emulsion.

Pseudocolour holograms

If you are interested in colour for its own sake, as distinct from natural colour, there is no need to buy or hire an expensive

125

argon–ion laser, nor to use two different types of plate: you can produce images in a whole range of colours using only your helium–neon laser. The colours are synthetic (hence the name), and to a large extent are under your own control. There are two basic methods. One uses multiple reference beams, and reconstruction beams of different colours; the other employs the 'rainbow' principle to reconstruct different image beams in different colours. In the first of these two systems, due to Mark Gaddis and David Welter, there are two object-illuminating beams, which we will call 'red' and 'green'. (Both are in fact red.) These are used separately to make two master holograms. In

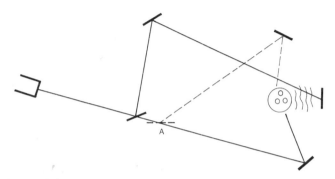

Pseudocolour hologram (multiple reconstruction beam): master hologram. **A** Removable mirror for changing direction of object illuminating beam between exposures. Two separate masters are made. For clarity the beams are shown unexpanded.

between the exposures, apart from changing the direction of the object-illuminating beam, you can remove some objects and add others, or spray areas of the objects with paints which alter their reflectance, such as green (to darken) or white or aluminium (to lighten). Having made your master holograms, you now make an image-plane hologram of the pseudoscopic real image, using the two master holograms to make the two separate exposures and switching the reference beam to the opposite side for the second, by inserting or removing a mirror.

To view the hologram you need two reconstruction beams, with filters respectively of red and green (or any other colour you may choose). These sources need not be collimated or coherent. The effect of changing the direction of illumination in the original will result in the image appearing to be lit by two lights, one red and one green, the region of overlap appearing yellow. The objects you remove will appear red, and those you add will appear green. If you apply paint, the objects you paint darker will be orange and those

126

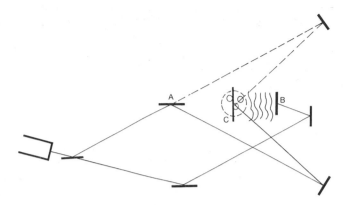

Final pseudocolour hologram. **A** Removable mirror (removed before second master exposure). **B** Master hologram. **C** Image in plane of film.

you paint lighter will be yellow-green. You can reconstruct using other colours: green and blue light will give a range of blue-greens; blue and red light will give purples and magentas. If you use complementary hues such as red and cyan (blue-green), the result will be in shades of red and cyan, with white as the mid-point.

The second method, first described by Poohsan Tamura, depends on the fact that in a rainbow hologram the colour of the image when viewed from a particular angle depends on the angle at which the reference (and reconstruction) beam strikes the film: the steeper the angle, the redder the hue.

The most spectacular results are those which reconstruct in the three primary hues, red, green and blue. You need to make three master holograms: you can change the direction of illumination, remove or add objects or paint objects between exposures. You can even make totally abstract colour compositions with crumpled white paper, changing it between exposures. For all three exposures the reference beam should be at 60° incidence at least, with a cylindrical collimating lens, as described on pages 101–2. Mount the masters between two glass plates with the 'red' master on the left, the green in the middle and the blue on the right. The angle between the red and the green, as seen from the final hologram film, should be 9°, and between the red and the blue 15°. If your final hologram film is 250 mm (10 in) from the masters, the spacing should be 39 mm (1.56 in) between the centre lines of the red and green masters, and 67 mm (2.64 in) between the red and the blue. These are the maximum allowable separations, and give the greatest colour saturation. Increasing the separation results in the blue image being reconstructed in invisible ultraviolet. Now you mask the masters down to 6 mm slits. The reference beam for the final hologram should be at 45° incidence.

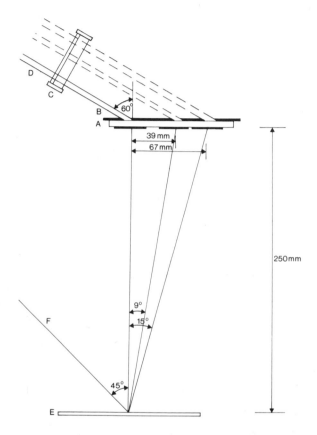

Pseudocolour hologram (single reconstruction beam). **A** Masters taped to glass, red on left, blue on right. **B** Opaque tape masks off unwanted areas. **C** Cylindrical collimating lens. **D** Master image reconstruction beam, moved between exposures. **E** Final hologram. **F** Reference beam (actually spread).

Expose one master at a time, moving the collimated object beam to cover each slit in turn. Bleach process the final hologram.

It is this last type of hologram which tends to be exploited to the greatest degree by holographers treating the medium as an art form. By combining a number of images at different distances from the final hologram it is possible to produce dramatic and beautiful effects.

128

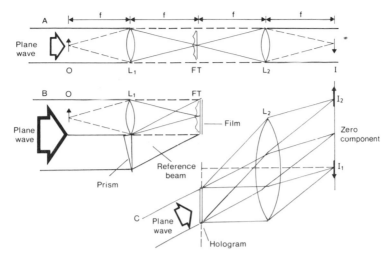

Principle of Fourier-transform holography. **A** Using arrangment for obtaining an optical image with two lenses L_1 and L_2 of the same focal length. **B** Making a hologram of the F–T using oblique reference beam. **C** Reconstruction. Two real images I_1 and I_2 are formed, one of them (I_1) in the position of the original image. Moving the hologram in its plane does not affect the position of the images.

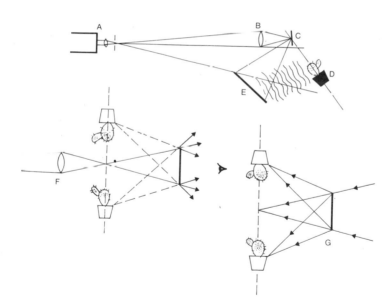

'Lensless' F–T holography. **A** Laser with beam spreader. **B** Lens. **C** Mirror at principal focus of lens. **D** Object. **E** Film, in plane parallel to plane containing object and principal focus of lens. **F** Reconstruction of virtual images using diverging beam. **G** Real-image reconstruction using converging beam. The focal length of lens **B** is immaterial unless non-holographic film is to be used, in which case then it should be quite long, and the object situated as close as possible to the mirror.

129

One final type of hologram is called a *Fourier-transform hologram*. This is the hologram of the diffraction pattern of an object formed at the principal focus of a lens. To produce an image for viewing it requires a lens of the same focal length, and it gives two images in the same plane, one of which is inverted. The depth of field is small. Fourier-transform holograms, first described by George Stroke, are easy to make, and do not require holographic emulsions. There are two basic arrangements: one has the lens in the object beam; the other has it in the reference beam.

Fourier-transform holograms are not particularly interesting aesthetically, though they have a number of fascinating scientific applications in such fields as pattern recognition and the improvement of photographic images (see Plates). For a fuller discussion the reader is referred to the standard textbooks (Appendix 5).

Part 3
Holography tomorrow

Has holography a future?

Where technology is concerned, it can be very unwise to speculate on the future. Many well-known figures have in the past made rash predictions, and have later looked very foolish. Holography is a young technology; much younger than many others such as photography and electronics, which are still developing, and, indeed, finding new uses every day. But holography, in a way, is different. The theoretical limitations are well understood; the range of possible holographic formats was described some years ago; their possible applications have been considered, and in most cases tried out and either accepted or rejected. Some of the applications await new advances in technology: for example, the development of an X-ray laser for diagnostic work; or of a coherent electron beam for holographic electron microscopy; high-quality readily-erasable thermoplastics film for computer software; and high-sensitivity photochromic glass for storage of data in the form of volume holograms.

One advance that is so far a pipe-dream is the making of holograms by white light. Of course, it is theoretically, as well as practically, impossible to use daylight or any other continuous spectrum. Yet it is possible to record a reflection hologram using three or more wavelengths simultaneously. Using a multi-wavelength source, with as many as ten wavelengths, we have a close approximation to white light, and this could lead to greatly improved colour holograms. It is theoretically possible to make integral holograms without a laser, using an ordinary mercury arc, as the thickness of the transparency is less than the coherence length of the light.

Holographic movies and television

A good deal of journalism about holography in the lay Press is nearer science-fiction than science. Space-fiction comic strip writers have had a ball with holography, with heroes projecting a holographic doppelgänger of themselves across the galaxy to do some cosmic cleaning-up job while they themselves remained behind. Science-fiction authors, too, have produced some odd ideas. A perennial favourite is holographic movies, with nubile images indulging in 3D strip shows. By now you will appreciate that although ciné-holography is feasible, and, indeed, has been achieved in several ways, the very limited 'window' of a hologram makes its viewing very much of a one-person exercise in voyeurism.

The most obvious method of ciné-holography would seem to be to record holograms with a pulsed laser at 16 or 24 per second. However, as holograms give a same-size image, 16 mm, or even 35 mm, film would not give much parallax except for tiny objects very close to the film, and only one person could view at a time. Multiplexed holograms seem a better bet, but a stereoscopic film projected onto a lenticular screen produces the same effect, and for a larger audience. Another suggestion which has been tried out is the recording of large numbers of images on a thick-emulsion volume hologram form. This could be played back by moving the reconstruction beam according to a preset sequence. However, even if advances in technology make it possible to store 1000 images in a single film, this represents only about 1 minute's viewing, with a very restricted 'window'. Because Fourier-transform holograms can be moved around without the image changing its position, the principle has been suggested as a basis for continuously moving film ciné-holography with wide films; a further idea is to sacrifice vertical parallax by making successive holograms as narrow horizontal strips on film, thus allowing projection with wide film running at quite a low speed (16 or 24 strips per second). The main drawback is still the small 'window' and limited image size; there have been suggestions for projection of real images using a screen of deformable thermoplastics material changing its structure 24 times a second in response to a flying-spot electron beam. This last suggestion would lend itself to holographic television, though there are theoretical as well as practical difficulties. The limited-window problem can be solved by hybrid techniques such as integral holography, though parallax has to be sacrificed in favour of movement, leaving one no better off than viewing the original movie.

134

The real objection lies not in the feasibility of holographic cinema, however, but in its relevance. Most of the realism of the cinema comes from the movement, not from parallax effects, as indicated by the continued failure of stereoscopic movies to arouse more than a flicker of interest among the general public. In any case, the movie fan would probably prefer his vast screens and surround-sound and our voyeur the real thing in a real nightclub!

Another science-fiction favourite is wall-to-wall holographic television. Well, wall-to-wall television is certainly no great problem: television projectors already have that capability. Holographic television is another matter. Although it is theoretically possible to transmit a hologram via a television channel, the huge information content necessitates a bandwidth about a million times greater than is available – and that is for a single hologram, not 25 a second. By substantially reducing the information content (beginning with removing vertical parallax) it is just possible to transmit holographic information within a television bandwidth; but, as suggested earlier, the method of display presents both theoretical and practical problems; again, there already exist simpler methods of achieving a stereoscopic image. As with the cinema, it seems fairly safe to predict that future developments will be in the direction of horizontal-parallax pictures by optical means.

Holography and perception

Revolutionary scientific and technological innovations always attract a kind of lunatic fringe which is prepared to extrapolate what it thinks is the essence of the discovery to produce all sorts of wild implications. One has only to think of the wave of charlatanry that followed Louis Mesmer's discovery of hypnotism, or Volta's of electricity. Similarly, advances in technology often produce a wake of popular fallacies as a result of only partial comprehension of the principles involved. Not many years ago it was commonplace to hear people complaining that their wristwatches kept stopping because they had too much electricity in their bodies; there are even now people who wear copper bracelets because of some fancied electromagnetic effect. In its early days, holography also attracted its share of cranks prepared to use the principle as an explanation of every aspect of behaviour from the migration of salmon to the pattern of scales on a butterfly's wing. Once the shouting had died down, however, two interesting facets of the implications of this new knowledge emerged. One was that holography can, indeed, sometimes offer the most feasible explanation of some aspects of behaviour. For example, it now seems reasonably certain that some species of dolphin and whale literally 'sound out' their surroundings by emitting pulses of ultrasonic radiation. They possess a large organ in their head, called the melon, which appears to function as some kind of detector. The conditions within the melon are virtually identical with those for producing an acoustic volume hologram, one which would be reformed with each pulse emitted by the creature. The direction-finding apparatus appears to depend more on subtle differences between the phase of the echo of its cry as received at each eardrum than on time-differences. Bats, which navigate and find their prey by ultrasound, also have a suitable 'receptor' in their brains. Even fish have a sense organ along their body line which appears to respond to pressure waves in a manner not dissimilar to that of an ultrasonic hologram.

The importance of these theories, however, is not so much that they can be proved to be true (or false) but that they can exist at all. Before holography they could simply not have been conceived.

There is another interesting aspect to the concept of the hologram. Once the principles of holography have been fully grasped, it becomes possible to employ these concepts in cognitive psychology, as the basis of a model for understanding aspects of cognition such as perception, recognition and recall. One can, for example, model cognition as a holographic store built up from perceptions from all angles; of recognition as a pattern-matching process akin to holographic character-recognition; and of recall as achieved by an appropriate stimulus analogous to the way an image is reconstructed from a volume hologram by the 'stimulus' of a light beam of the exact wavelength and from the exact direction required. It is important, however, to realize, as we stressed in the early chapters, that a model is not a description of how cognition and memory actually work: it is a metaphor which helps us to organize our observations and make predictions. One such prediction based on the holographic model would be that damage to part of the brain would in general not erase a certain group of memories, but simply impair or blur all memories to a greater or lesser extent; this fits the observations. But the subject is a complicated one, and some schools of thought hold that any use of models in this context is at best misleading and at worst stultifying.

136

Holography as an art medium

There seems little doubt that of all the possible applications of holography, two have a fairly secure future. The first of these is in metrology. There is no denying the importance of holographic interferometry in the science of measurement; and holography is now firmly established in the metrology laboratory. The second is as an art form. Although holographic images have obvious applications in advertising and display as well as in visual education, as an aesthetic medium they stand or fall on the basis of what they have to offer the viewer as an artistic experience. It is easy to allow oneself to be misled by one's feelings when looking at a display hologram. The gee-whiz nature of some exhibitions of holography can readily lure one into acceptance of the notion that a full-size three-dimensional image of a hollow pipe or a telephone set in itself provides an aesthetic experience. Unfortunately, the excitement is mainly in the novelty of the experience; and as the novelty wears off, so does the excitement. The really pertinent question is: is it possible to produce a genuinely creative hologram? If a hologram by its very nature is a facsimile of a three-dimensional object, can looking at it be an experience that is any different aesthetically from looking at the original object? If the object is, say, a piece of sculpture, why not just look at the sculpture?

It may perhaps be helpful to look at a parallel example: pictorial photography. In its early days photography evoked the same kind of wonder that holography does today; and the notorious phrase 'From today painting is dead' echoed round the Paris salons.

As it turned out, the impact of photography on painting was just the opposite, forcing painters into new and exciting ways of self-expression. It was photography, rather than painting, that dragged its feet. For its first five or so decades, its practitioners aimed only at reproducing subject matter as faithfully as possible, often going to the lengths of setting up elaborate tableaux that replicated well-known realist paintings. The shortcomings of the photographic emulsion, such as its insensitivity to red light, its graininess, non-linear tone-scale and lack of colour, were to be overcome as far as possible, and for the rest made fairly unobtrusive; technique was directed chiefly to this end. Consequently, a photograph was thought of only as a record of a scene, the aesthetic contribution being limited to the arrangement of the subject matter, and to some extent the lighting. As a result, very few critics afforded photography the status of an art form; many,

137

indeed, argued that the phrase 'creative photography' was a contradiction in terms.

Then, in the 1920s, things began to change. Photographers began looking for ways of exploiting the very shortcomings older generations had been at pains to conceal. They began to deliberately introduce grain, high contrast and tonal distortion, strange perspectives with wide-angle lenses and swing-backs, multiple exposures and Sabattier-effect reversals – and all in a genuinely creative way. A new generation with, among others, Man Ray, Bill Brandt and Moholy-Nagy emerged, and showed the critics that it was possible to be genuinely creative in terms not only of subject matter, but of the photographic process itself.

Now, with the arrival of holography, we are back at the beginning again, artistically speaking, with a totally new kind of representational medium, and we hear the same argument from the critics: holography, by its very nature, cannot create, only replicate. At the same time its practitioners complain of the inadequacies of the medium: of speckle; pseudoscopy; astigmatism; lack of colour, or distorted colour; 'wiped' images due to movement during exposure; interference fringes; volume effects and so on. But these are precisely the things the creative holographer can exploit.

There is so far only a handful of workers in the field of genuinely creative holography, though in this respect it would be grossly unfair to exclude the realist school of professional display holographers such as Nick Phillips of Holoco. With his associates he has been producing large holograms for exhibitions for some years; and though impressive chiefly for their size and astonishingly high technical quality, many of these, such as simulated moonscapes, stage settings and science-fiction film sets, show a feeling for lighting and display that go far beyond mere reproduction. Among the pioneers in purely creative holography are Margaret Benyon and Harriet Casdin-Silver, both of whom have held numerous successful exhibitions of their own work. One or two painters (notably Salvador Dali) also experimented with holography fairly early in its history.

Margaret Benyon has made double exposures in which one object is seen through another or occupying the same space, and volume effects, as when separate images appear successively as the hologram is turned through an angle. She has also experimented with objects which cause multiple light reflections, or cast shadows on the holographic plate so that these shadows are visible in ordinary light; with 'non-holograms' where an object that has moved during the exposure forms a three-dimensional 'black hole' and with holograms included as part of larger creative concepts. At the time of writing she is working at the Australian National University in Canberra.

In the United States, Harriet Casdin-Silver, has created completely abstract patterns in three dimensions; twisted patterns of light. Much of her work, however, exploits the *trompe-l'oeil* possibilities of high-quality real-image holograms with somewhat surrealistic arrangements of objects and startling lighting effects. Recently, she has been exploiting the possibilities of multiple-image integral holograms; she has also been working on optical systems which pre-distort the image in order to counteract the distortion inherent in integral holograms. She runs a course in creative holography at the Centre for Advanced Visual Studies at the Massachusetts Institute of Technology as part of a higher-degree programme and as a unit of work for undergraduates and special students.

Sam Morree and Dan Schweitzer have produced some very beautiful white-light transmission holograms where a change of viewpoint in the vertical plane causes the picture, as well as the colour, to change. They use realistic images, but in an impressionistic way; for example using the rainbow effect to suggest a transition of daylight into sunset. Rudie Berkhout works with completely abstract shapes in pseudocolour. Bill Molteni specializes in abstract compositions produced by diffraction gratings, and has worked on the elimination of vertical bars from holographic stereograms. He has produced successful stereograms of outdoor subjects. Amy Greenfield works in the field of integral 'movies', one example of which depicts a woman rolling in waves. Rick Silbermann's work exploits the anomalies of the holographic process, producing reflection holograms of pseudoscopic shadow-grams. Randy James produces multi-image white-light transmission holograms in colour; and Anait Stephens also works in the abstract, but with reflection holograms. Peter Claudius has made a name for himself with his integral holograms of erotic subjects. In Sweden, Karl Frederik Reuterswald is a professional artist who uses several holographic images in juxtaposition to produce particular effects, and Hans Bjelkhagen, who works mainly with laser transmission holograms, is also experimenting with new ways of producing white-light transmision holograms.

Conclusion

As yet, only a handful of professionals are exploring the limits of creative holography. Some of the most stimulating work has come from those who are already pursuing a career as artists in other media. Others have entered the creative field via scientific and technological paths. Much excellent work has resulted from collaborations between artists and technologists. Between them, these professional artists of holography are forging a new art form; one which breaks new aesthetic barriers with each new creative achievement.

So where does the amateur fit in? Well, in the field of creative pictorial photography, many of the greatest names belonged – and still belong – to those who began as amateurs. Indeed, many of the discoveries that revolutionized the art of the photographic image came from people with little or no formal training. In holography, there are still too few amateurs; far too few. We need original, creative thinking from those who have nothing to lose by being nonconformist, who make images just for the love of it – for that is what 'amateur' means. The techniques are not difficult, and the results are rewarding, as I hope this book has shown. Creative holography is still young and, as an amateur, you will be in a position to make a unique contribution to this new and immensely exciting art form.

Acknowledgements

Numerous people have freely given information and advice during the compilation of this book. I should particularly like to express my gratitude to Nick Phillips of Holoco, who made numerous valuable suggestions at an early stage; Bob Mooney of The Polytechnic, Wolverhampton, who helped me with equipment and accommodation; Duncan Croucher of Agfa-Gevaert; Michael Wenyon of Laser Focus; John Brown of Holographic Developments; Eve Ritscher, who was a mine of information on professional holographic artists; Ralph Wuerker of TRW Systems; George Stroke of Michigan University; Kaveh Bazargan of Imperial College, London and P. Hariharan of CSIRO, Sydney, Australia, who supplied me with technical information and photographs; Keith Hodgkinson of the Open University; Stephen Benton of the Polaroid Corporation; Margaret Benyon of Canberra University; Harriet Casdin-Silver of MIT, and, of course, my editor Rex Hayman, ARPS. Formula No. 3 on page 150 is reprinted with permission from 'Photographic Handling' by Stephen A. Benton, *Handbook of Optical Holography*, edited by H. J. Caulfield, Copyright Academic Press, Inc., 1979. Most of all I should like to thank Nigel Abraham of See-3 Limited for his enthusiastic support of the project and for the immense amount of information he obtained on my behalf, without which this book would have been far less complete.

Wolverhampton
August 1979

141

Appendix 1 Why holograms work

Light detectors such as photographic emulsions respond only to time-averaged intensity, which is numerically equal to half the square of the amplitude. We have seen, in an intuitive way, in the early pages of Chapter 3, that adding a second beam of light to the first is a way of coding phase as well as amplitude information into a hologram. But how does it encode it? And how does the illumination of the hologram with the reconstruction beam decode it?

If you have done a little trigonometry at some time you should be able to follow the argument easily enough. The only relationship you need to remember is the one that goes:

$$2 \cos X \cos Y = \cos (X + Y) + \cos (X - Y)$$

If we represent the reference and object wavefronts by U_{REF} and U_{OBJ} respectively, then for incoherent light the combined intensity at any point is simply $\frac{1}{2}(U_{REF}^2 + U_{OBJ}^2)$. However, if the beams are mutually coherent the resultant intensity is given by:

$$I = \frac{1}{2}(U_{REF} + U_{OBJ})^2 = \frac{1}{2}U_{REF}^2 + \frac{1}{2}U_{OBJ}^2 + U_{REF}U_{OBJ}$$

which is quite different, as the third term in the expression contains phase information. We can see this if we write U in the usual notation for a travelling wave:

$$U_{REF} = A_{REF}\cos(2\pi ft + \phi_{REF}(x,y))$$

$$\text{and } U_{OBJ} = A_{OBJ}\cos(2\pi ft + \phi_{OBJ}(x,y))$$

where A represents peak amplitude; f is the frequency; t is the time variable; and $\phi(x,y)$ is the phase of the wave at any point (x,y) on the emulsion. As the reference beam is unmodulated, A_{REF} is constant, and ϕ_{REF} is either constant (if the beam is perpendicular to the emulsion) or is directly proportional to x and y (if it is oblique). Because of diffraction by the object, A_{OBJ} and ϕ_{OBJ} will both be complicated functions of x and y.

Writing out the equations in full we have:

$$I = A_{REF}^2\cos^2(2\pi ft + \phi_{REF}) + A_{OBJ}^2\cos^2(2\pi ft + \phi_{OBJ}) + 2A_{REF}A_{OBJ}\cos(2\pi ft + \phi_{REF})\cos(2\pi ft + \phi_{OBJ})$$

Using the identity $2\cos X\cos Y = \cos(X + Y) + \cos(X - Y)$ to expand the third term gives:

$$I = A_{REF}^2\cos^2(2\pi ft + \phi_{REF}) + A_{OBJ}^2\cos^2(2\pi ft + \phi_{OBJ}) + A_{REF}A_{OBJ}\cos(4\pi ft + \phi_{REF} + \phi_{OBJ}) + A_{REF}A_{OBJ}\cos(\phi_{REF} - \phi_{OBJ})$$

Now this is time-averaged. The time-average of a $\cos^2$ function is $\frac{1}{2}A^2$, and the time-average of any cosine function is 0. However, the last term in the equation does not vary with time (there is no t present). Hence the equation simplifies to:

$$I = \tfrac{1}{2}A_{REF}^2 + \tfrac{1}{2}A_{OBJ}^2 + A_{REF}A_{OBJ}\cos(\phi_{REF} - \phi_{OBJ})$$

which, as can be seen, contains information on A_{OBJ}, the amplitude, and ϕ_{OBJ}, the phase, of the object wave.

A correctly exposed and processed emulsion has an amplitude transmittance which is directly proportional to the intensity of the illuminating wave. If we illuminate the hologram with a reconstruction wave U_{REC} identical with the original reference wave:

$$U_{REC} = A_{REC}\cos(2\pi ft + \phi_{REF})$$

and the amplitude transmittance of the hologram is proportional to I, then the image wave U_{IM} will be proportional to $U_{REC}I$. Thus:

$$U_{IM} = A_{REC}\cos(2\pi ft + \phi_{REF})(\tfrac{1}{2}A_{REF}^2 + \tfrac{1}{2}A_{OBJ}^2 + A_{REF}A_{OBJ}\cos(\phi_{REF} - \phi_{OBJ}))$$

$$= \tfrac{1}{2}U_{REC}(A_{REF}^2 + A_{OBJ}^2) + A_{REC}A_{REF}A_{OBJ}(\cos(2\pi ft + \phi_{REF})\cos(\phi_{REF} - \phi_{OBJ}))$$

Again using the identity $2\cos X\cos Y = \cos(X + Y) + \cos(X - Y)$, we have:

$$U_{IM} = \tfrac{1}{2}U_{REC}(A_{REF}^2 + A_{OBJ}^2)$$
$$+ \tfrac{1}{2}A_{REC}A_{REF}A_{OBJ}\cos(2\pi ft + \phi_{OBJ})$$
$$+ \tfrac{1}{2}A_{REC}A_{REF}A_{OBJ}\cos(2\pi ft + 2\phi_{REF} - \phi_{OBJ})$$

The first term is simply the reconstruction beam, attenuated. The second term is the object wave multiplied by a constant, and this means that we have succeeded not only in encoding the wave in the hologram, but also in recreating it by means of the reconstruction beam. The third term is also similar to the object wave, but has an extra term $2\phi_{REF}$ which changes the direction of the wave by an angle $2\phi_{REF}$, forming the real image. The negative sign before the ϕ_{OBJ} term reverses the phase of the wavefront, inverting its shape. This accounts for the pseudoscopic nature of the real image.

Appendix 2 Some useful addresses

Suppliers of lasers

Spectra-Physics (UK) Ltd
Laser Products Division
17 Brick Knoll Park
St Albans
Herts AL1 5UF
(0727)30131

Spectra-Physics
Laser Products Division
1250 West Middlefield Road
Mountain View
California 94042
(415)961–2550

Coherent (UK) Ltd
13 The Mall
Bar Hill
Cambridge CB4 4BH
(0223)68501

Coherent
Laser Division
3210 Porter Drive
Palo Alto
California 94304
(415)493–2111

Laser Lines Ltd
19 West Bar
Banbury
Oxon OX16 9SA
(0295)57581

Jodon Engineering
Associates Inc
145 Enterprise Drive
Ann Arbor
Michigan 48103

Rofin Ltd
Winslade House
Egham Hill
Egham
Surrey TW20 0AZ
(07843)7541

Edmund Scientific Co
Edscorp Building
Barrington
New Jersey 08007
(609)547–3488

Laser kits and electronic components

Maplin Electronics
Supplies Ltd
PO Box 3
Rayleigh
Essex SS6 8LR
(0702)554000

Optical components for holography

Offord Scientific
Equipment Ltd
113 Lavender Hill
Tonbridge
Kent TN9 2AY
(0732)364002

Edmund Scientific Co
(see above)

Ealing Beck Ltd
Greycaine Road
Watford WD2 4PW
(0923)42261

Microscope objectives

Pyser Ltd
Fircroft Way
Edenbridge
Kent TN8 6HA
(0732)864111

Edmund Scientific Co
(see above)

Pinholes

Graticules Ltd
Sovereign Way
Tonbridge
Kent TN9 1RN
(0732)359061

Ernest F Fullam Inc
PO Box 444
Schenectady
New York 12301
(518)715–5533

Films, plates and chemicals for holography

Agfa-Gevaert (UK) Ltd
NDT Sales Dept
Great West Road
Brentford
Middx
(01)560–2131

Agfa-Gevaert (US) Ltd
275 North Street
Teterboro
New Jersey 07608

Kodak Ltd
PO Box 66
Hemel Hempstead
Herts HP1 1JU
(0442)61122

Metax Ltd
PO Box 315
Gladstone Road
Croydon
Surrey CR9 2BL

Eastman Kodak Co
Rochester
New York 14650

Edmund Scientific Co
(see above)

Newport Research Group
18235 Mt Baldy Circle
Fountain Valley
California 92708

Holograms

See-3 (Holograms) Ltd
13 Bovingdon Road
London SW6 2AP
(01) 736 0076

Holographic
Developments Ltd
10 Marshalsea Road
London SE1 1HL
(01)403–1717

Holoco
Shepperton Studio Centre
Squires Bridge Road
Shepperton
Middx

Holex Corp
2544 West Main Street
Norristown
Pennsylvania 19401

The Museum of Holography
11 Mercer Street
New York
New York 10013

Edmund Scientific Co
(see above)

Appendix 3 Holographic emulsion data and processing formulae

Both Eastman Kodak and Agfa-Gevaert produce several emulsions suitable for helium–neon laser holography. At the time of writing it is not easy to obtain Kodak holographic films and plates in the UK. They are distributed in the United States by the Newport Research Group, whose UK outlet is Metax Ltd; and if you particularly want to work with Kodak materials you may have to persuade your retailer to order from them, or deal direct if you need large quantities. Agfa-Gevaert materials are equally good, and are available in a larger range of sizes. They can be obtained from professional photographic dealers: in case of difficulty you should contact Agfa-Gevaert direct; the relevant addresses are given in Appendix 2.

The Kodak emulsions are SO–173 film and 120–02 plates (for transmission and reflection holograms) and 649F (for reflection holograms with higher resolution). 649F is about 2½ times slower than SO–173. There is also a much faster emulsion, SO–253, which is suitable for transmission holograms, though its resolving power is not adequate for reflection holograms. All these materials are available in 4 × 5 in size, and SO–173 is also available in 35 mm size. Agfa-Gevaert make two emulsions, 10E75 for transmission holograms and 8E75HD, which is about 5 times as slow, for reflection holograms. These are available in 70 mm unperforated rolls, 4 × 5 in and 8 × 10 in sheets and in rolls 114 cm × 35 m. 10E75 is also available in 35 mm. Plates are available in 4 × 5 and 8 × 10 in size, and in larger sizes up to 100 cm square by special order. There is also an emulsion, 8E56, which is sensitive to green and blue, and suitable for argon–ion

148

laser holography. Agfa-Gevaert emulsions are known as 'Scientia' emulsions in some countries.

Both Kodak and Agfa emulsions are normally supplied without antihalation backing. For best results holograms (except master holograms for the various two-stage techniques) should be bleach-processed. This not only increases the brightness of the image considerably, but helps to minimize the effects of unfiltered or 'dirty' beams. Agfa-Gevaert market a bleach-processing kit suitable both for transmission and reflection holograms; unless you have a rooted objection to buying ready-made processing chemicals you would be well advised to use this. However, as in amateur photography, many people like to experiment with their own formulae; there is no denying that there is still much to be learnt in the field of holographic processing chemistry. For those who prefer to make up their own formulae, a selection of the best to date follows.

Developers

There are four very satisfactory developers. The first is the simplest, but also the most expensive. The second is simple and cheap, but will fetch the skin off your fingers unless you use rubber gloves: it also softens the emulsion and makes it very delicate. The third is similar but more complicated and somewhat better; and the fourth is a fairly traditional formula.

For both reflection and transmission holograms

1 Tetenal Neofin Blue used undiluted. Development time 5 min at or slightly below 18°C (65°F). Improved contrast results from adding 0.3 grams (g) per litre antifogging agent (benzotriazole) and 120 g per litre sodium metaborate (Kodalk). A substitute for Neofin Blue which is very nearly as good, and very much less expensive, is marketed by Agfa-Gevaert under the code number GP61:

2 *Solution A:*

Water	350 ml
Ascorbic acid	10 g
Sodium sulphite	5 g
Water to make	500 ml

Solution B:

Water	350 ml
Potassium hydroxide	10 g
Water to make	500 ml

149

Weigh out the potassium hydroxide as quickly as you can, as it absorbs water from the atmosphere and thus steadily gets heavier. For the same reason you need to keep it in an airtight container (this also applies to sodium hydroxide). Mix the two solutions together immediately before use. Use at full strength. The mixed solutions do not keep. Develop for 5–6 min at not more than 18°C (65°F).

3 A slightly more complicated developer, due to Stephen Benton and reproduced here by courtesy of Academic Press Ltd, gives a somewhat brighter image. Mix the ingredients in the order given. This developer does not keep.

Water	700 ml
Disodium hydrogen phosphate (Na_2HPO_4)	28 g
Sodium hydroxide	12 g
l-Ascorbic acid	18 g
Phenidone-A	0.5 g
Ammonium thiocyanate	0.5 g
Water to make	1000 ml

This developer is suitable for reflection holograms on Agfa-Gevaert 8E75HD. For Kodak plates and film the quantity of ammonium thiocyanate should be doubled. When used for transmission holograms the phenidone should be omitted.

4	Water	700 ml
	Metol	6 g
	Hydroquinone	7 g
	Sodium sulphite	30 g
	Sodium carbonate	60 g
	Potassium bromide	2 g
	Phenidone	0.8 g
	Disodium EDTA*	1 g
	Water to make	1000 ml

Development time 2 min at 20°C (68°F). This developer is unsuitable for reflection holograms made with a helium–neon laser, as it tans and swells the gelatin, resulting in an infrared reconstruction; but it is excellent for argon–ion laser holograms.

* Sold by garden shops under the brand name of Sequestrene. Not necessary if you use de-ionized or distilled water.

Kodak D-8 and D-19 are commercially available developers which are also quite satisfactory. Other workers have had success with tanning developers such as pyro-metol and pyro-catechol, the formulae for which can be found in the photographic literature; but this technique is unreliable for reflection holograms as the gelatin becomes cross-linked in a swollen condition, with the result that the reconstruction wavelength is shifted so far into the deep red that the image becomes all but invisible. However, it produces a pronounced relief effect which has been used to produce copies of transmission holograms on metal foil by pressure contact.

Fixing (transmission holograms)

Master holograms should be fixed and washed in the same way as photographic negatives. About 3 min in Kodak Rapid Fixer (without hardener) diluted 1:3 is sufficient. Wash the fixed film for 10 min in running water. Add a drop or two of wetting agent to the final rinse and hang the film up to dry. In hard-water districts it is a good idea to wipe off surplus moisture gently with a cellulose sponge.

Bleaches

There are several methods of bleaching, one of each type being represented here. Direct bleaches convert the silver image back into silver bromide; reversal bleaches remove the silver, leaving the undeveloped silver bromide in the emulsion. Total bleaches remove all the silver and silver bromide, leaving only a stain image.

1 Agfa-Gevaert GP-432. This is probably the best of all the bleaches. It is a direct bleach which is suitable only for films which have been very fully exposed, to a density of 3 or more.

Potassium bromide	30 g
Boric acid	1.5 g
Water to make	1000 ml
Just before use, add:	
Parabenzoquinone	2g/1000 ml

After development wash the film for 5 min in running water, then immerse it in the bleach bath. Leave the film in the solution 1 min after the bleaching appears complete, then wash for 5 min and dry without heat.

151

NOTE: Although parabenzoquinone is no more toxic than many other chemicals regularly used in photographic processing, it is dangerous because it is in the form of a very fine powder, which you can easily inhale if you weigh it out carelessly. It can also be absorbed through the skin when in solution, so wear rubber gloves when you are carrying out this part of the process.

This bleach does not remove any of the silver from the emulsion. It relies on the redistribution that has occurred during development, which is the reason for requiring such a heavy exposure.

2 Agfa-Gevaert GP-431. A type of bleach suitable for films exposed to a mean density of about 2 is based on ferric nitrate. This is a direct bleach, which takes place after fixing and washing.

Ferric nitrate crystals	150 g
Potassium bromide	30 g
Water to make	1000 ml

Add
Phenosafranine   0.3 g
(dissolved separately in methanol)

This is diluted 1:4 for use. Leave the film in the solution for 1 min after the bleaching appears complete, then wash and dry as before.

3 This bleach can be used with either unfixed or fixed films. In the former case it is a reversal bleach; in the latter, it is a total bleach.

Potassium dichromate	2 g
Potassium bromide	30 g
Water to make	1000 ml

Add slowly, with stirring:
Sulphuric acid, conc.   2 ml

Add:
Phenosafranine   0.3 g (not necessary with fixed films)
(dissolved separately in methanol)

This bleach is suitable for fixed transmission holograms of density 1.5 or somewhat less, or for unfixed reflection holograms of density 2 or greater.

Most long-standing holographers seem to have their own pet bleach. Some of these contain highly poisonous chemicals such as mercuric chloride, or irritants such as bromine. The main problem associated with the processing of reflection holograms is distortion of the gelatin, which usually takes the form of shrinkage in thickness accompanied by an increase in visual noise and loss of contrast. To restore a shrunken emulsion to its original thickness, immerse it (after drying) in:

Water	700 ml
D-sorbitol	100 g
Wetting agent	5 ml
Water to make	1000 ml

Soak the film for up to 30 min in this solution, then remove it. Carefully remove surplus solution with a cellulose sponge, and allow to dry naturally without washing. If you find you have overdone the swelling and driven the image into the infrared, you can restore it with a 2 min wash in plain water.

NOTE: Previous workers have recommended a 3 per cent solution of triethanolamine with Kodak Photo-Flo. However, this combination resensitizes the emulsion to light, and it should not be used for bleached holograms containing residual silver bromide.

Appendix 4   Making a holographic collimator

Any transmission hologram will produce a real image if illuminated with a reconstruction beam which is the precise opposite of the original reference beam. This technique exploits the phenomenon. If you make a hologram of a point light source using a collimated reference beam, then illuminate it with a collimated reconstruction beam from the opposite side, this will reconstruct a real image of the point. Effectively, the hologram is acting like a lens with a focal length equal to the distance of the original point source from it. Conversely, if you illuminate the hologram with light from the original point light source it will reconstruct the collimated beam. The hologram is what is called a zone plate; it behaves like a collimating lens in every respect except that the emergent beam is along a different axis from the incident beam. It follows the usual rules of geometrical optics, and is eminently suitable as a collimating lens in all two-stage holographic techniques.

It is not necessary to have a collimated reference beam to make this holographic collimator, as the same effect can be produced by using another point source at a different distance from the 'object' source. If the two point sources are at distances D_1 and D_2 respectively from the film, the focal length F of the holographic zone plate is given by the formula:

$$\frac{1}{F} = \frac{1}{D_1} - \frac{1}{D_2}$$

Provided D_1 and D_2 are large compared with the diameter of the zone plate. Let us suppose that you want a zone plate with a focal

154

length of 100 cm. If you choose suitable values for D_1 and D_2 (say 60 and 150 cm) so that:

$$\frac{1}{F} = \frac{1}{60} - \frac{1}{150} \left(= \frac{1}{100} \right)$$

you will finish up with a zone plate which behaves like a collimating lens of the focal length you require.

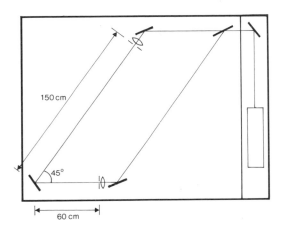

Sandbox layout for 100 cm holographic collimator. For clarity the beams are shown as if unspread.

The two beams should be at about 45° to one another and their angles of incidence should be equal, i.e. 22.5°. The two beams should be of roughly the same intensity, though you may get better diffraction efficiency if the intensity of the longer beam is slightly higher. Expose the film to give a density suitable for bleach processing.

To use the collimator you simply illuminate it using a point source such as a microscope objective and spatial filter from a

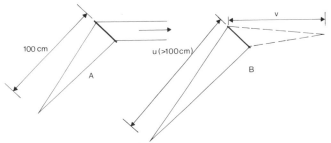

Using the holographic collimator. **A** Producing a parallel reference beam.
B Producing a converging reference beam.

distance of 100 cm and at the correct angle of incidence for reconstruction. By increasing the distance of the source from the zone plate you can get a converging beam if you require one. The beam will converge towards a point given by the relationship

$$\frac{1}{v} = \frac{1}{F} - \frac{1}{u}$$

where u and v are the distances respectively of the object and image from the zone plate, and F is its focal length.

Appendix 5 Books and courses on holography

Howard M. Smith: *Principles of Holography* (2nd edn), John Wiley & Sons, 1975.

Good historical introduction. The mathematical theory is separate from the descriptive text. Processing methods are discussed in detail. There is a full discussion of non-silver processes, including photopolymers, photoresists, dichromated gelatin and thermoplastics. There is an excellent chapter on scientific applications of holography.

George W. Stroke: *An Introduction to Coherent Optics and Holography* (2nd edn), Academic Press, 1969.

A rigorous mathematical treatment of the subject. Stroke was one of the pioneers of holography, and discovered the principle of Fourier-transform holography. There is a non-mathematical appendix by I. P. Nalimov on the applications of holography (up to 1967), and Gabor's historic papers of 1948 and 1951, in which he first described holography, are reprinted in facsimile.

Robert J. Collier, Christoph B. Burckhardt and Lawrence H. Lin: *Optical Holography*, Academic Press, 1971.

Every conceivable aspect of optical holography, including the theoretical and practical limitations of all processes, is covered in depth, and there is also a chapter on computer-generated holograms. There is a comprehensive mathematical treatment of the relevant optical theory.

Takanori Okoshi: *Three-dimensional Imaging Techniques*, Academic Press, 1976.

This book is devoted to all forms of stereoscopic and full-parallax imagery. It includes discussions of the physiology and psychology of depth perception, lens-sheet pictures, projection displays, and holography. The mathematical treatment is in terms of wavefront theory for holography and geometrical optics for the remainder. There is a comprehensive section on holography in colour, and a detailed analysis of methods of information reduction to enable the transmission of holograms by television.

Matt Lehmann: *Holography: Theory and Practice*, Focal Press, 1970.
Intended primarily for research workers, this book deals mainly with the practical aspects of holography in optics laboratories. There is some basic information on photographic optics and sensitometry, and the mathematical explanations of diffraction and interference are at a comparatively simple level.

Michael Wenyon: *Understanding Holography*, David & Charles, 1978.
A remarkably lucid account of the principles of lasers and holography, non-mathematical and aimed at the intelligent layman. Contains a chapter on the industrial and military uses of lasers. There is a very full discussion on the possibilities of holography in display and advertising, and as an art form. There is a final chapter devoted to making your own holograms using an anti-vibration table made from a paving slab.

Ronald M. Benrey: *How to Build a Low-cost Laser*, Hayden Book Co, 1974.
The first one-third of the book is devoted to building your own 1 mW laser. The design of the power supply is somewhat out of date; but Maplin Electronic Supplies Ltd (see Appendix 2) can supply PCBs and components for an updated version. The remainder of the book is devoted to optical experiments and small-scale holography using a home-made optical framework.

Christopher Outwater and Eric van Hamersveld: *A Guide to Practical Holography*, Pentangle Press, 1974.
The practical section, which takes up rather less than half of this short guide, is very much down to earth. There are instructions for building a heavy optical table from timber and steel sheet. The magnetic component holders demand a certain amount of metal-working ability. There are clear diagrams, with distances and angles shown, for transmission and single-beam reflection holography, and some rather less precise suggestions for other setups. The first part of the book deals

158

with coherence, lasers and principles of holography. It is enthusiastic though not always clear, and contains a number of errors.

George Dowbenko: *Homegrown Holography*, American Photographic Book Publishing Co, 1978.

Intended for amateurs. The second part of the book shows how to build a sandbox with bricks and concrete, and gives step-by-step instructions for most types of hologram. The first part of the book, dealing with principles of optics, lasers and holography, is much less satisfactory. It uses a mixture of current and obsolete units, and contains many errors. The hand-drawn diagrams are whimsical and often misleading. The practical section is somewhat better.

H. J. Caulfield: *Handbook of Optical Holography*, Academic Press (printing).

Eve Ritscher: *Holography as an Art-form* (in preparation).

Comparatively few books have been written on holography in recent years, but a great many papers and articles have been published. Research papers appear from time to time in the scientific journals dealing with optics and telecommunications. The Museum of Holography (see Appendix 2) publishes a newsletter called *holosphere* which gives details of recent developments in holography, and which is available by subscription. Amateur photographic periodicals sometimes carry articles on holography, but these are not always reliable. *New Scientist* (weekly) and *Scientific American* (monthly) carry news and articles on advances in holography, among other scientific matters.

Courses

The Open University: Course ST291: Images and Information (revised 1977). Open University Press.

A second-level half-credit undergraduate course, dealing with principles of wave optics, spatial filtering and holography, and adopting a Fourier approach to imaging. There are case-studies involving astronomy, spacecraft, electron microscopy, ultrasonic imaging and computerized tomography. The home experiment kit includes a small laser, a 35 mm holographic camera and a small optical bench. The course is backed up by radio and television programmes and computer-based tutorials. Prerequisite knowledge of mathematics is trigonometrical functions only: there is no calculus. Units 3–10 (three books, available separately from OU stockists) deal with coherence, diffraction, lenses, spatial filtering and holography.

Many universities and polytechnics offer undergraduate and postgraduate courses in scientific holography, but in the UK, there are as yet (1979) no courses in creative holography, though there are plans to run one at Goldsmiths College, London, and another at South Hill Park Arts Centre, Bracknell, Berks. In the United States, there are a number of schools of creative holography, but apart from those quoted in the text, they tend to operate sporadically. It is suggested that readers in the United States contact the Museum of Holography (see Appendix 2) for details of current courses.

Appendix 6 Glossary of terms

Achromatic

Describes a lens or objective which is free from dispersion (q.v.). When used in describing holograms it means simply 'uncoloured'.

Additive colour synthesis

A method of producing an image in natural colours by mixing lights of different hues, typically red, green and blue.

Amplitude

In general terms, the maximum value of the displacement of a point on a wave from its mean value. Instantaneous amplitude is the actual displacement at any given instant. Time-averaged amplitude is the average value of the instantaneous amplitude over a number of cycles.

Amplitude hologram

A hologram in which the information is coded in the form of variations in transmittance (cf. phase hologram).

Astigmatism

The horizontal and vertical aspects of the holographic image are formed in different planes. Note that this is not the same as the definition applicable in photographic optics.

Beamsplitter	A partially reflecting mirror or prism which divides the intensities of the transmitted and reflected beams in a fixed ratio.
Blu-tack	The trade name of a synthetic putty adhesive marketed by Bostik Ltd and sold in the USA under the name of Superstuff. It is particularly useful as a temporary support for optical components.
Bragg diffraction	The principle by which a stack of parallel reflecting surfaces (or zones of alternating high and low refractive index) will reflect a beam if, and only if, the reflected wavefronts are of the appropriate wavelength and orientation to produce constructive interference.
Brewster angle	The angle of incidence of a light beam such that the reflected and refracted beams are mutually perpendicular. At this angle (typically about 55°) the reflected beam is totally polarized in a plane perpendicular to the plane containing the incident and reflected beams.
Brewster window	An optical window offset so that the light beam concerned is incident on it at the Brewster angle. Such a window has zero reflectance for a beam which is polarized in a plane perpendicular to the plane of the window.
Carrier beam	Another name for a reference beam (q.v.), drawn from radio communications technology.
Coherence length	The maximum difference in length of optical path between two light beams originating from the same source that still allows the formation of visible interference fringes.
Collimated beam	A beam of light which neither converges nor diverges.

162

Colour temperature	A description of the spectral energy distribution of a light source in terms of the absolute temperature at which a perfectly radiating body would produce the same energy distribution.
Continuous-wave (CW) laser	A laser emitting a light beam the intensity of which does not vary with time (cf. Pulsed laser).
Cosine grating	A one-dimensional grating with a transmittance that varies cosinusoidally with distance.
Cosine wave	A wave whose instantaneous amplitude varies cosinusoidally.
Cosinusoidal	Fluctuating in the same manner as the function $y = A \cos x$, where x and y are the variables and A is the peak amplitude. The fluctuations are identical with the sinusoid $y = A \sin x$, but are shifted in phase by one-quarter of a cycle. The cosine function is an *even* function, that is, it is symmetrical about the y-axis.
Cross-talk	The production of spurious images in a colour or multiple-image hologram.
Deconvolution	The removal of unwanted elements of a picture (e.g. raster lines) by modification of its optical Fourier transform.
Denisyuk hologram	A single-beam hologram in which the object is on the opposite side of the holographic film to the reference beam. Thus the reference beam, on passing through the emulsion, becomes the object-illuminating beam on the other side.
Density	In photographic sensitometry, the common logarithm of the reciprocal of the transmittance. It is properly called 'optical density', and in colorimetry is known as 'absorbence'.
Depth of field	In photography, the distance between the farthest and nearest points which

give an image of acceptable definition at a given lens aperture. In holography, the depth of field is the space which will record on a hologram. It is dependent upon the coherence length of the light source and the holographic configuration.

Diffraction

The change in direction of a wavefront encountering an object. Strictly, the term subsumes both reflection and refraction, though for practical purposes these are usually treated as distinct phenomena.

Diffraction efficiency

The proportion of diffracted to incident light in a hologram.

Diffraction grating

A one-dimensional grid of ruled lines, used for dispersing a beam of light into a spectrum.

Direct-beam hologram

Any holographic configuration not employing a beamsplitter.

Dispersion

The separation of a polychromatic beam into its component wavelengths.

Electric vector

The electrical component of electromagnetic radiation, vibrating perpendicular to the magnetic vector. Linearly polarized light is said to have the plane of polarization that applies to the electric vector.

Electromagnetic radiation

The propagation through space of the effects of a fluctuating electromagnetic disturbance.

Excited state

In the Bohr atom, a state in which one or more electrons are in a state of higher energy than the ground state.

Fabry-Pérot etalon

An optical cavity formed by two very accurately parallel glass–air surfaces, used in Q-switched lasers (q.v.). A beam of light entering the etalon undergoes multiple reflections which interfere constructively only for a wavelength which is an integral submultiple of the distance between the faces.

Focused-image hologram	An image-plane hologram (q.v.) where the image is formed by a lens.
Fourier-transform hologram	Strictly, a hologram made in the principal focal plane of a lens, but also applied to one made using a spherical reference wavefront originating from a point in the object plane. In both cases the reconstruction leads to a pair of images, one erect and the other inverted.
Frequency	Used without qualification, the term is synonymous with temporal frequency (q.v.).
Fraunhofer hologram	Another name for a Fourier-transform hologram (q.v.).
Fresnel hologram	The usual type of hologram, in which the object is close to the film.
Fresnel lens	A lens which is cut back in concentric steps to reduce its thickness.
Fresnel zone plate	A pattern of concentric rings, alternately transparent and opaque, their radii chosen such that on passing a beam of monochromatic light through the plate the diffracted wavefront converges to a point. By reversing the direction of the light it is possible to produce a collimated beam from a point source. When made holographically, a zone plate has approximately cosinusoidal variations in transmittance, and is sometimes known as a Gabor zone plate.
Gabor hologram	An in-line hologram (q.v.).
Gas laser	A laser in which the pumping energy is supplied by an ionized gas.
Ground state	In the Bohr atom, the condition in which all the electrons are in their lowest-energy orbitals and the atom is in a stable energy state.

165

Hologram	A complex diffraction grating which, when illuminated appropriately, produces an image of an object, usually with full parallax. The grating is usually, though not invariably, produced as a silver image in a photographic emulsion, and is an interference pattern generated by the wavefront from the object (the object beam) and an unmodulated wavefront (the reference beam). The various types of hologram are listed individually.
Image beam	The diffracted beam leaving a hologram which reconstructs the object wavefront.
Image-plane hologram	A hologram in which the 'object' is a real image produced by a lens or another hologram. Depending on the position of the image relative to the plane of the film, the holographic image may be real, virtual or partly real and partly virtual.
In-line hologram	A hologram in which the reference beam passes through the object space.
Integral hologram	Also known as a multiplex hologram, this is a hybrid of holography and photography in which a series of narrow vertical strips of transparencies taken from adjacent viewpoints over an angle up to 360° is made into a cylindrical hologram.
Intensity	The square of amplitude (q.v.).
Interference	When two coherent wavefronts are superposed, their instantaneous amplitudes add at every point. If the amplitude of the resultant is greater than that of the component waves the interference is said to be constructive: if it is less, the interference is said to be destructive.
Interferogram	The record of an interference pattern produced holographically or by tradi-

	tional interferometric methods. Used for measuring very small deformations or disturbances.
Kelvin	The fundamental unit of temperature. 0 kelvins (0 K) = $-273.18°C$ (absolute zero). A change in absolute temperature of 1 kelvin is the same as 1°C. The unit is named after Lord Kelvin, a pioneer of thermodynamics.
Lippmann hologram	Another name for a reflection hologram (q.v.).
Michelson interferometer	A device based on interference of light beams, used for extremely accurate measurements of displacement.
Microwave holography	Any holographic technique using a coherent beam of microwaves, rather than light.
Monochromatic	Literally, single-coloured, that is, of a single wavelength. Often used loosely as a comparative term to describe the degree of temporal coherence (q.v.) of a source.
Monochromator	A device for filtering out light of unwanted wavelengths from an incoherent or partially coherent source.
Multiples and submultiples	When basic units of measurement are too large or too small for convenience in a particular situation, they may be multiplied or divided in steps of a thousand and given a prefix. Those used in this book are as follows:

Multiples	*Submultiples*
kilo (k)	milli (m)
× 1000	÷ 1000
mega (M)	micro (μ)
× 1 000 000	÷ 1 000 000
giga (G)	nano (n)
× 1 000 000 000	÷ 1 000 000 000

See also Scientific notation.

Note: In this system, which is internationally standard, the centimetre (1/100 metre) is a 'rogue' unit used as a matter of convenience in non-scientific measurements. It appears in this book where use of 'correct' units would be unnecessarily pedantic.

Neutral-density filter	A light filter which has the same transmittance for all wavelengths, used for attenuating a light beam by a known factor. The number given to a particular filter (e.g. 0.6 ND) represents its density (q.v.).
Object beam	In holography, the modulated beam formed by the light waves scattered or reflected by the object onto the film.
Object-illuminating beam	The unmodulated beam which illuminates the object.
Open-aperture hologram	An image-plane white-light transmission hologram made using the whole of the master hologram area (cf. rainbow hologram).
Optical cavity	Usually this term refers to the space between the mirrors of a laser (but see Fabry-Pérot etalon).
Orthoscopic image	An image which has correct parallax.
Oscillator	Any device which converts energy into an alternating electromagnetic field.
Parallax	The phenomenon whereby a different view of an object is obtained by changing the viewing position.
Partial coherence	In a beam of electromagnetic radiation, the property of the constituent photons remaining in phase over a short distance (typically of the order of 100 wavelengths).
Period	Used without qualification, the term is synonymous with temporal period (q.v.).

168

Phase	The relationship between the position of the crest of a wave and a given reference point. It is measured in radians (2π radians = 1 cycle), degrees, or fractions of a cycle.
Phase hologram	A hologram in which variations in transmittance are replaced by variations in refractive index or thickness.
Photon	The smallest amount of electromagnetic energy that can exist. In theories of coherence, a beam of electromagnetic radiation is considered as being made up of large numbers of photons. The temporal frequency of a photon is directly proportional to its energy.
Plane wave	The wave description of a collimated beam, the wavefronts being parallel planes perpendicular to the direction of propagation.
Population inversion	The condition of a substance containing atoms in two different energy states at an instant when there are more atoms in the higher energy state than in the lower.
Polarization	Electromagnetic radiation is said to be polarized (strictly, linearly polarized) when the electric vector vibrates in one plane only.
Principal focal plane	In geometric optics, the plane in which a beam of collimated light entering a lens converges to a point.
Product	In the context of Fourier-transform theory, the product of two functions is the result of multiplying their values together at every point.
Pseudoscopic image	A holographic image with reversed parallax.
Pulsed laser	A laser which emits radiation as a short burst.

Q-switching	A method of storing electromagnetic energy in the optical cavity of a pulsed laser until it has built up to a very high value, then releasing it in a burst of high intensity, high coherence and short duration.
Quasi-monochromatic	Used to describe a light source (such as a laser) which emits only a very narrow band of wavelengths.
Rainbow hologram	Also known as a Benton hologram after its inventor, this is a white-light transmission hologram produced in two stages, the second hologram being made using the pseudoscopic real image of the first as object. Only a narrow horizontal strip of the master hologram is illuminated, so that the vertical parallax is eliminated and its place taken by an optically generated diffraction grating. The image, viewed by white light, is seen in one or other of the spectral hues depending on the angle of viewing.
Raster	The grid of lines which carry the picture information in a television display.
Real image	An image formed by light waves which actually pass through the image space. A real image can be recorded directly on photographic film.
Real-time interferometry	In holography, a technique whereby an object is superposed on its own virtual image. Any movement or distortion of the object is contoured by interference fringes.
Reconstruction beam	The unmodulated beam which is directed at a hologram to recreate the object wavefront.
Reference beam	The unmodulated beam which is directed at the holographic film when making a hologram. Also called the carrier beam.

Reflected-beam hologram	A hologram made with an object beam reflected from the object (cf. Transmitted-beam hologram).
Reflection hologram	Any hologram intended for viewing by reflected light.
Refractive index	The ratio of the speed of light in empty space to that in a given optical medium. In a phase hologram it is the phase delay factor per unit thickness referred to empty space.
Resolving power	The highest spatial frequency a given recording medium can record. Its reciprocal (the lowest spatial period) is known as the resolution.
Scientific notation	All numbers are written with the decimal point immediately following the first digit, the number being then multiplied by 10 raised to the appropriate power (which will be negative for numbers between 0 and 1). Thus the speed of light (299 700 000 metres per second) is written 2.977×10^8 metres per second, and the wavelength of helium–neon laser light (632.8 nanometres or 0.000 000 632 8 metres) is written 6.328×10^{-7} metres.
Sideband hologram	Another name, drawn from a telecommunications analogy, for an off-axis hologram, i.e. any hologram where the reference and object beams reach the film along different paths (cf. In-line hologram).
Single-beam hologram	A hologram in which the same beam serves as both reference beam and object illuminating beam.
Slit hologram	A more accurate name for a 'rainbow' hologram (q.v.).
Spatial coherence	The degree to which a beam of light is collimated or appears to have originated

	from a true point (for a collimated beam this point is at infinity).
Spatial filtering	Properly called spatial frequency filtering, this is a method of modifying image quality by placing opaque, partly transparent or phase-modifying stops in various regions of the principal focal plane of a lens.
Spatial frequency	The number of cycles of a repeated pattern in a given distance.
Spatial period	The reciprocal of spatial frequency, i.e. the distance between points of repetition.
Speckle	The grainy appearance of an object illuminated by coherent light. It is caused by interference between waves diffracted by minute surface roughnesses.
Stereogram	An image (generally photographic) which creates a three-dimensional effect by presenting different views of an object to the two eyes of an observer.
Stimulated emission	The emission of a photon by an excited atom when struck by a photon coming from a further atom previously in the same state of excitation. The second photon is of the same frequency as the first; in addition it travels in the same direction and has the same phase.
Submultiples	See Multiples and submultiples.
Subtractive colour synthesis	The principle used in making photographs in colour. Each of three layers of emulsion, sensitive respectively to red, green and blue, is developed to a positive in dyes which remove the correct amount of the respective primary hues from white light.
Temporal coherence	The degree to which the constituent photons in a beam of light have the same wavelength (frequency) and phase.

Temporal frequency	The number of cycles of a repetitive function occurring in a given time. Inversely proportional to wavelength. Usually known simply as 'frequency'.
Temporal period	The reciprocal of temporal frequency, i.e. the duration of one complete cycle. Usually referred to simply as 'period'.
Time-averaged interferogram	An interferogram (q.v.) showing fringes caused by interference between wavefronts produced by a vibrating object at the two extremes of movement, where it is momentarily stationary.
Transmission hologram	Any hologram which is to be viewed by transmitted light.
Transmittance	The proportion of light transmitted by an optical medium to that incident on it, expressed as a decimal or a percentage. In coherent optics it is necessary to specify whether amplitude or intensity transmittance is being considered.
Transmitted-beam hologram	A hologram made with the object beam transmitted through the object (cf. Reflected-beam hologram).
Ultrasonic hologram	Any hologram made using ultra-high-frequency sound energy.
Virtual image	An image generated by light waves which do not actually pass through the image space. Such an image can be seen, but cannot be recorded directly on a photographic emulsion.
Volume hologram	A hologram made on an emulsion many wavelengths thick. The image is visible only when the reconstruction beam is very accurately aligned.
Wavefront	The locus of points in a coherent light beam which are in the same phase.
Wavelength	The distance between adjacent wave-crests of a wave. It is inversely proportional to frequency, and to the

173

refractive index of the optical medium the wave is travelling in. In spite of the latter, the term 'wavelength' is used rather than frequency when discussing the behaviour of light, the figure used being the wavelength in empty space.

White-light transmission hologram A transmission hologram which can be viewed by white light.

Zone plate See Fresnel zone plate.

Index

Image,
  orthoscopic, 43, 66-8, 93, 97
  pseudoscopic, 43, 62-4, 68-9, 93-4
  real, 37, 63-4, 68-9
  virtual, 36, 63, 71
Incoherent light, 16
Ionization, 22
Interference, 32
  fringes, 25, 40, 42, 91, 112
  mirrors, 29, 42, 45
  patterns, 33, 34, 37, 51, 88
Interferometer steadiness test, 79-81
Isolation mounting, 51-2, 76-7

Laser, 15
  argon-ion, 31, 109, 115, 122
  buying, 49-50
  carbon dioxide, 31
  care of, 50
  continuous-wave (CW), 28
  dye, 31
  finding coherence length, 78
  helium-cadmium, 122
  helium-neon, 28, 29, 31, 50, 122
  low-power, 49, 68
  mirrors, 27, 29
  nitrogen, 31
  pulsed, 28, 110-11
  ruby, 26
  safety precautions, 50
  semiconductor, 31
  tubes, 50, 51
Latent image, 45
Lenses,
  glass, 97
  liquid filled, 97-9, 101-2, 104, 105
Light sources,
  incandescent, 15, 23
  incoherent, 16, 17
  gas discharge, 16-17
  partially coherent, 17, 44, 65
Line spectrum 17, 23

Mercury vapour lamp, 17, 18, 22, 39, 133
Michelson interferometer, 80
Mirror,
  ball bearing, 86-7
  cylindrical, 104
  dichroic, 112
  front surface, 61
  interference, 28, 42, 45
  roller bearing, 104
Monochromator 25

Neutral-density (ND) filter 52, 55, 60

Object,
  beam, 35
  illuminating beam, 38
Optical,
  bench, 54, 58-60
  cavity, 27, 29, 110
  coating of mirrors, 29
  path length, 40, 44, 81
Orthoscopic real image, 43, 70-1, 96

Parallax, 33, 36, 43, 44, 93, 96, 100, 119
Phase, 17, 19, 25, 32-5,
Photometer, 89-90
Photon, 13, 16-19, 20-8
Physical development, 45
Polarization, 29-31, 49, 82
Population inversion, 26, 28
Processing of films and plates, 55, 68, 149-53
Pumping energy, 28

Q-switching, 110
Quasi-monochromatic light, 27, 52

Reconstruction beam, 36
Reference beam, 35
Relay mirror, 78

Sandbox,
  construction, 74-6
  optical component holders, 78-9
  sand, 77
  siting, 73-4
  steadiness test, 79-81
  supports, 77
Sodium lamp 22, 64
Spatial filter, 49, 83-4
  pinhole alignment, 85-6
  pinhole diameter, 84
  pinhole manufacture, 84-5
Speckle, 51, 104, 107
Stereogram, 33, 118
Stimulated emission, 25-8

Uneven illumination, 57, 80

Wavefronts, 32
Wavelength, 14

Zone plate, 154